A Brief Atlas of My Inner World

Jen Baranovic

Cover photo © 2021 by Jen Baranovic

A Brief Atlas of My Inner World/ Jen Baranovic

Composed April 2017- June 2018
ISBN 13: 978-1722921002
ISBN 10: 1722921005

Dedicated to all the selves I have written and left behind in these places, in these pages. Thank you.

Denver

City culture in the US has a phenomenon called Happy Hour, typically multiple hours before dinner time, in which people in business-casual livery flock to restaurants to drink off the stress of office job trivialities. I come today because I have no better way to spend a lonely, rainy afternoon than chasing my wandering thoughts down rabbit holes, writing, and drinking champagne. My trip here wasn't intended to be *this* trip, this alone trip sitting in day-lit bars and ambling around inside my own head, but I'm making do with what it is. The contrast of the engrossing reunion experience I envisaged so many times seems discordant against this isolating alternate reality, so I mentally include myself here like I'm at a people-zoo. At least it keeps me from wondering what you are doing on the island instead of being here. And it keeps me from thinking about the complicated chain of events that have unfolded in the past few years, culminating now in the weirdness of concurrently healing and running away from myself.

I observe the domesticated, polished paragons of civilization in these low-key post-gentrification venues with a half-hearted smile, knowing how close I came to becoming one of them. These are a strange breed of animals... frenzied business warriors living for stolen laughter over loudly clinking stemware. In calendared weekends, they escape the downtown to recreate in a nearby nature, still peopled and prescribed. I wonder if they are happy. I wonder if I would have been.

I decide that from an outside perspective, I appear to fit here well enough, but I am both absorbing and transcending this environment. It's like I need to keep one foot out the door in case this place tricks me into falling into one of its cages, never again to be released to the wild. In any case, I enjoy this observational exercise deeply in my flush-faced effervescence. I'm connected to everyone by proximity and warm transactional relationships, and yet I'm delightfully separate. I'm aware of the limitations of this peculiar slice of life, a culture which this cosmopolitan species of human may only perceive as consensual reality. But there is a trapped-ness of their aggregate psyche that is palpable in the shadows here despite the splendor of modernity and superficial vibrance. Happy Hour is likely a medical necessity if you carve out a rat-race existence in such a place.

As much as I've traveled in other countries, open-mindedly meandering around cities and villages, I have only just started doing the same in the US: going to places and abandoning my expectations, sampling whatever narrative a place offers up, daydreaming stories about people's lives there, becoming absorbed in microculture. Maybe I just thoughtlessly assumed that I knew how places in the States would feel, anticipating the same monotony of chain restaurants and dress codes and predictable tropes. Or maybe while I habitually backpack alone internationally, I am almost never a table-for-one on my home turf. It's not lost on me that my mood overprints the scene. In any case, with new eyes and the oddness of my own current company, I sit here as an anthropologist, an observer of both other people and of myself. I couldn't feel more outside of both worlds.

I dawdle through all my favorite neighborhoods downtown, and I realize that I feel different being in a city. I am an energetic chameleon. I can usually adapt to the pace of life, the codes of transportation and transaction, the costume and character that qualifies one's entry into various subsets, the *feel* of things. It usually takes me a day or two to acclimate and then the essence of a place seeps into my cells, a river that sweeps me along with it. I am soluble in my surroundings.

I feel like I belong everywhere and nowhere lately, moving fluidly between different realities. What job would I take if I lived here? I imagine myself as an artist, renting a flat above the pulsing heart of the central district. Or maybe I'd have stayed in the sciences and would have worked my way up in some big company. So many simultaneous versions of my life unfold in my mind, the multiverse expounding itself in a bloom of dimensional awareness: who I might have been, who I could still be, who I am, the choices we make that lead us down the path of a possible self. I indulge this for awhile, casually trying on lives like I'm trying on shirts in a dressing room.

My attention snags and I'm pulled back to some higher form of myself, the one that doesn't change so freely. There is a part of me that sits back and watches me play my existential identity game. It dawns on me that the "me" that I become in this place is not the most evolved edition, despite my buzzed contentment with the introspective curiosity it inspires. The city form of me is a consumer. I'm materialistic here, and enthralled with it until I take a step back. I indulge my senses with the finest cuisine and mixology creations. I think more about image here than I typically do: I covet clothing and unique haircuts and tattoos as I walk. I browse stores that are a tactile menagerie of softness, texture, form. I wonder more about how I represent my person in this visually stunning sea of creatively expressed individuals. Am I really the person that someone would see me as? How can I better decorate myself to express my essence? Everything is commoditized. I'm shoveling money into the desperate furniture fire I've built to stave off the cold of some nondescript emptiness here.

My thoughts on image are part art and freedom, part toxic capitalist entrapment, and it confounds me on a level. To buy the adorning things that showcase my perception of self is to put that part of me up for sale. There are so very many places poised to feed my starved self-esteem. I decide it's better to abandon my fucks about perfecting some deliberate persona. Still, the bandages are peeled back from the wound of self-concept amid a barrage of perfectly defined characters.

The desire to eat and drink and touch and wear and dance with and make love to everything here, as though *this* is how you could ultimately experience the true spirit of a place, remains eddying hungrily in the peripheral. But the deeper part of me touches the edge of the enormous loneliness, picking at it like the dewy paper label on a bottle. *That* part of me remains dissatisfied at the ease of my seduction by specious charms. I have another champagne to shut that part of me up, which explains everything I'm feeling in a way I couldn't have said better.

I force my return to gratitude, swimming back up from the depths. I decide to embrace the moment and this place for what it is, what it brings out in me, and the weird and complex facets of being human that drinking alone can highlight. I am all of a sudden happy to be here, happy to know more of the world than just here, happy to be alive.

You cross my mind again, and once more I feel paradoxically alone and connected. There is nothing I can do but let go, over and over and over. There is nothing to do but be part of where I am, and part of the Everything where you also are. I'm half drunk and it suits me wholly. I sit and write, sometimes to you and sometimes about you, sometimes about him, them, others. Three haunt me from the ethers, none of them Real here. It's a tangle, and so am I. I look at the page and decide to leave it as it is: confusing but honest. And in that way, perfect.

I leave the restaurant and the skies open up. It pours rain for the whole ten block walk back to my hotel. People skitter like ants, cowering under umbrellas and darting from awning to awning. I don't try to fight getting wet. It's a beautiful reminder that even here among the thick of glass and concrete and civilization, nature finds us and touches us, soaks us if we let it.

I let it.

I look for the solution here
In this parallel universe
version of me
I look for signs as I wander the city alone
But whatever I seek on the other side of the door
Every door
Is some magic that refuses to be shaken loose from
mundane and predictable scenes,
Can't be unlocked from shops or strangers,
Rooftop bars or dive hotels.
The barrier that prevents entry to something Real
Is still right here,
Within.

Cursed to search the world over
For some externality that
will cure the disconnect,
Some leverage for me to see the self I abandon.
I meet the eyes of people on the street
and they are wells of their deepest losses
and mirrors to my own:
The family I never started
The loves I never finished
The lives I dead-ended
The Mother who never calls.

Is it any doubt that I can't heal these wounds
Emptying bottles and scrolling through my phone
Alone in my high-rise
Alone in my head
or sitting in the window seat of countless coffee shops
Waiting for the plot to change

I could have lived a thousand different lives by now
Which one could have stopped the
emptiness from spilling over,
Which one could have stopped the seeking?

My consumer-self remains unsatisfied because there is always more to want. I am voracious and inconsolable. The paralysis of too many choices pairs well with the emptiness of the law of diminishing returns. I'm insatiable, or not. I'm numb and bored and can't tell. I'm agitated. And I'm being separated from my money quickly, a fact that makes me nauseous when I think of the unfulfilling things I'm trading it for. I'm hungry but indifferent and choose the hotel restaurant. The prices are extortionist. I'm too lazy to leave so I order water and soup, and "the most introverted table available".

The bisque arrives at my dark booth and is unremarkable, probably poisoned by my mood. I mentally muse that I should have asked for a half portion of something else instead, as it seems like the kind of place where people can just make ridiculous requests. *I'd like peppermint tea steeped in melted glacial ice and served at 101 degrees, please.* People here can get whatever they want except for what they really want.

My soul is rapacious, and my body feels fat. Clearly my wellbeing practices entirely fall apart in the charade of first-world hotel life, and I make a vow in this very moment, *ok after the eleven dollar cheesecake*, to strengthen them.

I oscillate dramatically here between resigned acceptance and despairing uneasiness. I long to be a centered, radiant goddess that exudes peacefulness and patience, the kind that uses the hotel gym or runs in parks, makes friends on the bus, or walks to a Whole Foods. I miss how I felt in Thailand. But here, I have overstayed and I am wounded, extrasensory empathic, scattered, diseased. I feel miserably alone. I miss my tea rituals, my own clean cooking, meditating in the sun of my tiny living room surrounded by my cat and potted plants. I miss gardening and dirty hands and simplicity. I've never been in a place with as many people that don't really meet or talk to each other. They stick to their pre-established groups like they're in glass bubbles... which is novel for a few hours of people-watching, but awful for real life. It's odd that I am longing for home. So often when I'm away, I'm exuberant and living in the moment. Right now, whether it's my disconnection within this place or within myself, I am so, *so* happy to be heading back tomorrow.

Yesterday I asked an Uber driver what he missed the most about his home country, Senegal.

"*Everything*," he replied in accented perfect English, his baritone thick with longing. "People are sooo nice there. Here... it's hard, it is stressful. But it's a land of opportunity."

I thought a lot about that in the night. It's easy to see that we trade community and connection for "opportunity": money. Those of us who are born here don't usually knowingly make this exchange, it's built into our lives to the point that we don't even see it. There is no tribe to choose instead. The US is the land where money is god but there's no kinship or singing in our churches. *iPhones are our altars*, I think, laughing at myself as I type this on the Notes app. I reflect on the vapid and self-centered values we have compared to other cultures I've witnessed... the injustice of privilege and wealth disparities; happiness or the endless pursuit of it, and what that has come to mean. But the cackling of boisterous mid-forties white middle-class restaurant patrons keeps me from pulling complete sentences out of abstraction. I couldn't have chosen a place more different from how I feel than this.

Don't go back to sleep, the Sufi poet Rumi said. I can feel myself, in general, on a path to healing after a few bad years of drama and emotional abuse. I am becoming whole again on the other side, but it is taking *so much time*. There is still confusion, angst, and misplaced defiance. Sometimes I truly don't know whether I am making decisions from a place of residual trauma or if I am through it, choosing freely. The disquiet of not belonging here infiltrates the experience now, and it unmasks a remaining issue: my difficulty feeling belonging *anywhere*, wherever I am. This feeling has increased with my self work, I think. It's as though the more I heal, the more I reveal the remaining ugly unhealed parts of me. Feeling split undermines my center with insecurity, and all my old maladjusted coping mechanisms come out to play in the gap. And those, after days alone here, seem much harder to overcome. It's like I'm awakening only to find myself in a dirty bed in a lurid old bedroom, and I don't yet have the means to wash the sheets or paint the walls. Sometimes I get closer to undertaking those tasks, and sometimes, like now, I suffer for only being able to distract myself from the unpleasant reality, the enormity of work to be done.

Leveling up is exponentially more difficult as you get higher. In fact, the longer I stay in these environments the more I feel my progress waning. I pay my tab and retire to my room with renewed resolve. I am going to make a list of the qualities within myself that I want to develop and how to go about it, maybe a list of things that make me feel fulfilled in my life. I have to *do* something to appease my inner voice that is screaming: *Wake up, wake up! Stay awake here! Everything here is designed to deter you from honoring your wildness. If you become tame, you play by the rules of those that will sedate your soul.*

The Pillars of Balance
*Learning new things
*Being in new cultures
*Planning for future experiences
*Making new connections
*Cultivating existing connections
*Being in nature
*Exercising and becoming stronger
*Having fun/ playing
*Helping other people
*Preparing food and eating clean
*Creative expression
*Keeping a clean, aesthetic home space
*Having integrity in relationships
*Meditating
*Being present
*Money- being deliberate

The New Rules
-No alcohol for awhile, eat whole foods
-Take time to do things, no multitasking
-Don't trade money for unfulfilling things
-Don't accept invitations you don't want
-Self-care, especially when depressed
-Prioritize health, wellbeing, self-development above all else
-Stop negative self talk, start where you are
-Attempt better sleep and detox
-Extend rather than absorb energy
-Invite people to meals/ activities. Build connections
-When traveling, focus on health and rest, quality of
        experience
-Don't fall back on poor coping because of emotional state

What you want should not be
A comparison of the things available to you to have
What you want
Should be divined from sticks
cast on the banks of the river of the soul
Manifested from the ethers of dreams and
pulled though the portal.

Tell me what you create in your mind
Not what options lie before you
Tell me what keeps your heart unhinged,
What keeps you up and restless long nights,
What makes you bleed?

My therapist recently told me that I have impulse control issues and an addiction to passion.

Yes, I say. I starve in boredom. I need connection, vibrancy, novelty. I need to feel things intensely, tactilely. Travel and fall in love often. Make new friends, see new places, sing new songs. I'm ravenous for the kind of big, delicious life that can't wait to be devoured whole.

She informed me that these things are potentially symptoms of a problem. And if I don't treat my problem, it could escalate. The treatment is pharmaceuticals (because that is the treatment for everything in our culture). I'm dismissive of this because I don't believe in a chemical imbalance that no one can actually test me for. *That's probably the kind of denial that a mentally ill person would have*, I laugh to myself.

Society would have me mute my emotional range so that I obediently enjoy the flatness of belonging like a pet to a city or a man, so that I take comfort in routine and predictability. But I'm not out of my mind, I just *feel* a lot. And thus far, if I have indeed earned a label for my mental condition, my "hypomania" has taken the shape of indulgent, extraordinary interludes of creativity and energy, not a catalyst for regrets or upheaval.

Is my capacity for a broad experience really an affliction? Do I need meds to temper my cravings to live an electric life on the very cusp of discovery of Self and other? Of home and away? There is beauty in duality, in contrast, in living in the moment. *I'm healing from a catastrophic emotional trauma*, and I want to be allowed to take the time I need to rebalance. We are taught that if something feels bad we must change it immediately: fix it, or run away from it, or numb it. But what I want to do is understand *why* I feel with such intensity, *why* my range of moods swings so drastically sometimes. And I can't do that if I change it before I take a long, hard look at it.

I suspect that this is only a chapter in my recovery, that my emotions are as all over the map because I'm rewiring my brain, and I haven't stabilized the new pathways yet. There are times when I revert back to the damaged thinking of a manipulated person and times when I gloriously shed that baggage and step out into the light. *You know what else is bipolar? The sun, the moon, the earth,* I joke to myself. But really, *I know myself,* and I do *not* want pills for this.

If you ask me, an adventurous life is clearly the only medicine (says my hypomania?). My healing escalated when I started saying YES. Yes to travel, to therapy, to concerts, retreats, classes, friends, all of it. Yes I want to cut my hair, yes I want to skip work and go on a road trip, yes I want to stay up all night and throw rocks at the moon. Those "yeses" are rebuilding my will to live.

I forget the lows when I'm like this. *Ehh, the lows maybe need meds sometimes,* I concede. The bottom is beyond all comprehension unless you're there in it. It wasn't more than a few weeks ago that I sat in my bathroom floor with the doors locked and my phone turned off, lightly tracing the blue of my veins with a razor blade... thinking. I spent hours googling how to commit suicide without making a mess. I'm not sure I ever particularly want to die when it comes down to it, I just can't *feel at the right scale* in those moments, and entertaining thoughts of dying is like a reset to zero. I have had three such episodes in the last year. I can't currently access the thoughts of that dark, secret person within me capable of such things, and the memories play like an unsettling dream. *But what if this is just a thing I'm going through?* How will I ever heal *for real* if I don't face it straight on? It's been a relatively short amount of time since I fell into this particular episode of depression. The person that triggered my spiral (or perhaps shoved me into the void) is no longer in my life. I feel like I just need some time and all this will level out on its own.

I tell my therapist that I won't go on meds for being bipolar just yet. I've only just quit the SSRIs they gave me for my previous diagnosis of Major Depressive Disorder. I need time to try to work through this naturally. Moreover, I believe that the pills are *responsible* for making me hypomanic. I want time to get them out of my system completely. I want to test my thyroid, my vitamins, my gut health. I want to *heal* if I am sick, not just fight the symptoms! This is not just in my head. I think the new diagnosis is wrong, extreme. I want to exercise and eat clean and take supplements to see if I can balance my moods by healing my whole person. I want to know my baseline without meds to understand completely whether I want them or need them.

She gives me the side-eye but eventually concedes. She warns me that I need to get help immediately if I get to the point where I irrationally quit my job or run away from the life I've built. I promise to check in on my rationale and ask people close to me to help. *Asking others to validate my reality is absolutely terrifying during recovery from narcissistic abuse.* But the issue with having a mental affliction is that you can't trust that you can actually see out of it. I undermine my own instincts, both good and bad, if I become too fearful that some self-destructive version of me is making my decisions. Studying yourself with this degree of objectivity causes some decoupling from reality, no doubt. I've definitely been battling my tendency to self-medicate my sad little animal. I am as confused in my healing as I was in the throes of my previous despair. I don't trust anyone else, and I definitely don't trust myself at times. But meds don't fix trust.

Impetuously, or maybe in a moment of clarity, it crosses my mind that therapy is actually making me feel worse at the moment by transferring blame out of the system of my Self and onto my past and the label of a condition. I'm not a victim, and thinking about mental state this way seems to remove ownership. It destroys the agency and accountability I know I need to overcome this more organically.

I decide to give it some time and be patient with myself. If I'm truly ill, I've been like this long enough to get through what it throws at me. I recoil at the idea of meds, especially after my previous horrible experience, but I want to live my best life. I think my hang-up is that there is no scientific test for serotonin levels, there is no brain scan that tells me how I am damaged. No one has looked at my genes and pointed to the one that makes me different. There is only anecdotal evidence, my own shifty self-reporting, and other peoples' opinions. Even if I really am bipolar and I come to accept that, what does it even mean? We are all somewhere on these spectrums, and we also all evolve constantly. My tendencies and behaviors recently may have qualified me for the label, but they haven't always and may not always. How will we know if or when I'm better if I'm on meds?

When I was in college, I rented an apartment from a lady people told me was manic depressive. She was absolutely nuts: screaming rants, ratty uncombed hair, disappearing for days. All I knew about "manic depression" was that these types of people were high drama and clearly should be avoided. I smugly applaud myself for hiding my apparent disorder behind a successful career, travels, a social life, publishing books, a full calendar, and a semi-polished demeanor. Am I really one step away from falling apart like that? It seems impossible. Although I am not ashamed of the diagnosis, I think it's because I don't categorize myself with my own stereotype of it. I *can't* be bipolar if I don't *seem* bipolar on the outside, right?

Yet part of me knows that there is *something* wrong. After a few years of therapy, and going over the behavioral criteria with my cautious and sympathetic therapist, the diagnosis seems to fit *right now,* but not categorically. If I medicate, what I'm afraid of is still having depression but losing the highs- the only reward for being such a sensitive emotional creature. I ride these waves of energy into personal festivals of creativity and immersive feasts of sensory delights. These episodes are the only time I know for sure that I feel unadulterated joy. The thought of dulling them scares me. What can pull me out of the lows if not the highs? What about finding the real cause of my mood issue and curing *that* instead of guessing with these chemical cocktails with crazy side effects? If I get on pills will I ever get off of them, and is it worth it?

Everyone has opinions, and I view all of them skeptically. My acceptance of the diagnosis waffles back and forth. My intuition tells me that I can recover from this current state… not that I can intellectualize my way out of it, but that I can *feel* my way out if I take my time and start where I am. The only one with this mind and body is me, and if I can trust that I'm viewing myself clearly, surely I can find the passage through the darkness. I feel too much, and it's killing me or saving me. I put my money on the latter. Trusting my gut feels less dangerous than handing myself over to meds, but the margin is admittedly thin.

Remembering comes in waves
I cried again last night
and at first I was disappointed in myself
Crying for the loss of you, still.
But I sat with my thoughts
and I realized
I don't weep for missing you,
I don't want to speak to you or sleep next to you again,
I don't believe my life would be better with you.

I weep for my Self,
For those little memories of when I was lost and shattered
and confused by you,
twisted in guilt like everything was my fault.
I remember them over and over trying to
make sense of them;
I weep for all the times I didn't understand
and my heart was breaking in half for trying to
make things right,
make anything right,
when I didn't know what I'd done wrong.

I weep for the ways I am different now
and I'm not used to it and it's scary.

I mourn my losses in moments,
like a movie reel of the sad parts:
There I am standing with you
in front of the cathedral in Krakow
tears that won't stop pouring
because I saw that it was really over
but I knew there would be so many more hurts to bear
before it Felt over.
There I am sitting on my back step when you
told me on the phone that she's better than me
and my heart fell out of my chest into the gravel,
where I left it for days.
There I sit in a rental car in a hot parking lot
so happy that you answered my call

but you're telling me that you don't want to see me
while I'm in town.
And I die a little because I know I will never, ever
come back there for you again.
There I am driving away from you with my things,
There I am by the pool when you finally admitted your lie,
There I am when I still made love to you even though
I was sick inside with how you'd become.

There I am in my bedroom that didn't feel like mine
when you told me
I was too unsupportive of your new relationship
to be your friend
nevermind the constant comparisons,
your nonstop stories of her,
the hurt that I could bear up to a breaking point
nevermind whether you were a good friend to me...

Me, who was trying but suffering
and feeling lost and misunderstood and
so separate from reality.

That's the day I almost killed myself.

I ruminate over these memories all connected to you
But these are memories of Me.

I weep for the child inside me who clung to you
through the emotional abuse
because you were Everything,
and you loved me so well at first.
I weep for the innocent part of me that stubbornly refused
to see how poisonous you were
Because I really needed that love to be Real.

What is a family, and where can I get one?
There are people on dates here
And it makes me miss you.
Warmth and laughter,
The dark red of draperies and wine and upturned lips,
Candlelight glints on gold things.
It's reminiscent of late year holidays
And late night dinners across from you in Rome
And I'm lonely here without you
Without Any One.

Every person you've met has a version of you
recreated in their memory
A You they pull into life whenever they think of you
There you are,
frozen in time
with the things you affected
The things you said or did that left a mark
morphed into personality traits
they assign to you forever,
Assumptions about how this You
still interacts with the world, fixed
Immovable
We compartmentalize each other for the convenience
of self-preservation when love falls apart
or
confine each other to generalizations
while we ourselves are allowed to grow and
learn from mistakes.
My ego suffers from imagining how you think of me now
So I try to take the care to
Not distill you into the worst version
nor
exalt you to the best
I try to remember that I don't know you anymore.

I wonder if you offer me the same courtesy.

You were everything once
And now you're just
something that happened to me.

You feel like a movie I watched
Or a dream I had.
A memory is a strange thing
An essence of someone,
A construct
Confined to the past
while they flow forward in rivers unknown
The movie is over
I'm wide awake
And I remember loving a version of you
That neither of us will ever know again.

Peru

I feel myself having become a creature of indulgence, a vessel for sensory exploration. I crave only to move through this life on an ocean current of emotions and memories derived from rich experiences. I think I could never really have regrets-everything, even the hardest of things, has been right on time. It's a complex realization, but it's not dismissive. These full-caliber experiments in being alive are formative. I want to live the breadth of the thing. I want the self-awareness and development that *makes* a person, and that kind of growth doesn't come from intellectualizing life from outside of it or above it.

My sense of purpose feels derived from this idea now. I am here to feel deeply, and transmute that depth into love and art. I am here to sieve and extract the myths from which I'm woven; to take the infinite, that which lays the heart open or explodes it into space, and distill it into the discreteness of a narrative to be explored and savored, challenged and reframed. This is what I live for: to *feel* these things, face these challenges, pass these tests, to *level up*. I am here to heal.

I am adrift now, probably lost, I can't tell. But this is how I will reclaim myself, rediscovering who I am or reinventing her moment by moment. Whoever is left of me, I will find her here at sea. Surely she is out here living unhooked from all reference points, with no sight of the shoreline.

I've had a revelation, and I owe it to my therapist. I saw her recently between trips, and while I thought I was in for a quick tune-up, I boiled over with emotion and confoundment that had apparently waited to reveal itself in the safety of her lemongrass-scented office.

I thanked her for helping me through the superficial (albeit obliterating) crisis during breakup with P. I explained that now, on the other side of that massive drama, I'm left with the strange feeling of a bigger burden uncovered, a greater existential struggle. My regular life feels flat. I'm constantly at the mercy of my oscillating polarity. The highs come from writing, traveling, learning, excesses, crushes, running away. My lows are crippling, severe, self-loathing. They have the lead weight of all my failures coupled with the inability to repair any of it. Oddly, the pits of despair don't scare me like the emptiness I feel from the pedestrian domesticity that everyone else seems so bizarrely content with. I can't sit still. I can only feel in extremes right now.

She let me finish, then explained with vehemence that the existential crisis was the *only* issue, the true issue. P was a symptom. I'd chosen him because he'd made me feel something, he'd rescued me from my relative monotony.

I wasn't prepared for this framing, and it was a breakthrough. I was surprised that in all my churning introspection I hadn't seen it that way. There were at least a few moments I could remember amidst the turmoil with P that I'd sensed a simultaneous dilemma: episodes where his controlling nature pushed me further into the private reaches of myself, places out of his reach. And in that core, I felt deeper sufferings: lack of freedom, lack of feeling seen, and abysmal boredom for the mundane life he tried to corner me into after the excitement of an unconventional courtship.

Maybe the conflict with P was leverage for me to recognize the real illness of my soul: that restlessness that drives me toward impulsivity and passion, and at times recklessness when I can't assuage it otherwise. For so long, I've had no objectivity on my own nature, just survival reactions one after another. I am starting to believe the diagnosis, but I still can't tell if it's a shock reflex in the aftermath.

Now I sit with the great problem only, the real problem only, confined to its theoretical expression. P and I haven't spoken for most of a year. The restlessness was there before him, briefly hushed in my infatuation, then explosively there again.

"You *are* an existential crisis," my therapist remarked pointedly.

I'm half proud, half insulted by this, and thus far completely unable to change it. At least now I can see myself more impartially and am not just drowning in my circumstances... but it's hard to learn how to be a human again after the obliteration of identity I endured, having been crushed by the weight of an impossible relationship, a toxic man, and my own shades of codependence. I am recovering in waves, but the clarity is raw, sensitive to the touch. I can't be home too long, can't be hugged too hard.

I split my life in half for two years, and then it split me in half. Now in the fractured wreckage, I find moments of peace only in my extremes, only in the novelties that palliate the pain. It doesn't make sense right now to pray for mediocrity.

Stay with me
As I fumble toward cohesion,
Containment,
The elusive lucidity
Hanging with urgent weight
in the laden clouds of consciousness,
those portents that can only be read by
skryers and seekers,
But that thicken the air in
palpable inevitability.

Entropy emerging now
as evolution of identity
Retold by the expressions of the moon
Over and over
Until her moods move as tides in your soul:
I am becoming more.
I am becoming less.
I am beginning again.

Stay with me here
As I attempt to touch god
And write her down
Before she turns away.
What are we if not mirrors for those we see
When the identity of the Mother
Remains eclipsed
and void of reflection,
Light bouncing off the vacant faces of Other,
Unable to achieve Self.

I dreamt this place, but I didn't know it was Cusco.

I had arrived late in the evening and slept until the following afternoon, tucked snugly under heavy layers of hand-woven wool blankets. Mine was a small room with tall ceilings and sparse furniture, windowless except a small aperture on the door that opened to the stone-cool hollow of the courtyard. I had no idea what time of day it was until I cracked the door open. Even though the breakfast hour had long since passed, when I emerged like a bear from an overwinter cave, the middle-aged woman at the desk smiled broadly. She waived her hands and rushed over, warmly insisting that I sit for coca tea and a tamal. She ushered me to a white, cast iron cafe table and busied herself with the tasks, in grand hospitality. When she brought the tea, we chatted some in Spanish. She asked about my travels and marveled at the duration of the journey to arrive. I asked her what I should see in my few short days there. I liked her immediately, her genuine love for her guests was encompassing. The coca tea was important, she said, because the elevation could be a problem for travelers. She refilled my cup.

Jetlagged but ready to salvage the remaining daylight, I emerged from the wicket within the oversized blue door of the hostel and entered a new place, new pace of life. Dust and low sunlight danced down the narrow cobblestone street, and I followed it with no agenda. I wound my way down the steep hill past the crumbling brown stucco façades, and into an expansive fountained plaza lined with immense cathedrals. The pulse of syncopated pop music spewed from passing taxis, and friendly shop owners slowed the cadence of their Spanish to invite me in. I soon found myself in a sprawling labyrinth of markets and made a mental note to come back once I was acclimated. Today, I was just absorbing it all.

The city feels like a mishmash of Oaxaca and Rome, but edgier than both, spiked with overtones of its own indigenous, colonial, and expat overprints. There are bits that feel distinctly *Cusco*: ladies in traditional garb leading llamas through the alleyways, the immaculate Incan stonework interbedded with more modern structures, the explosion of taste and texture of ceviches and pisco sours. It is at once polished and impoverished, opulent and ancient. It feels grittier than I imagined, and for this I love it even more.

Even at first glance, there are hints of the thriving subcultures of healers and magic. Street signs openly boast ayahuasca ceremonies, drum circles, and reiki. There are Spanish classes, salsa dancing lessons, yoga studios, hikes and adventures awaiting. The restaurants have a broad range that includes vegan art dens, upscale cevicherias, single-table local dives, and the lively stalls of San Pedro market. It is at once catacombs and kantu flowers, complicated and beautiful.

It is a place that might be explored for a few days to form an impression, or lived in for longer to crack into the next layers of life. *I could stay awhile*, I think. But it's probably best that I have plans that keep me moving on. For now, I am enthralled. It's strange and complex here, and I am strange and complex. I'm dark, complicated, a little desperate, and on the cusp of both madness and brilliance. Cusco looks back and says, "Same."

He betrays me
So I punish him with a secret
And worse,
by slipping away enough
to want secrets
But I'll never win a game of indifference
His is absolute,
The core of his nature
And mine is a protest against the cruelty of being unloved
Unloved
Unloved.

A successful relationship or connection is one that wherever the rivers of passion, madness, and the messiness of life carry the individuals, they retain a admiration and respect for each other, and can wish the other well in all endeavors, including with other loves or chapters of life. I hold this space for several who can't be easily classified as merely a lover or partner or friend or ex... there is this beautiful *other* thing that defies a label. When this happens mutually, it makes my heart sing with hope for humanity, as though synapses of love connect like lightning across the clouds of the elevated. Love, defying time and space and convention. Love, prevailing against the chafe of norms, against those small-minded or primal impulses that cause us to hate or reject that which we can't own, conquer, or compartmentalize.

Maybe I transform these connections, these people, into abstractions so that they are mine to keep in the realm of my mind, and I haven't escaped those chains of romantic bondage after all. Or maybe some pathological part of me stubbornly refuses to let go, so I hold a delusion of continuation when a reconciled reality would prove it dead. But I truly don't think so. While I can postulate theories of how my belief in such transcendent loves could be erroneous, my heart and higher self know without doubt that these loves really do exist.

We do this with our dearest friends- love them forever and unconditionally- so it's not a far stretch to imagine we can be this way with lovers. I can't help but to have a thirst to find others that think this way. I live fully, in flow, in those shared moments of the creation of unity with and within another, the chemical reaction of language and touch, the eagerness to explore, to bridge, to open. These things come together in a cocktail whose intoxicating power is magnified exponentially compared to any one ingredient: a combustion of the senses.

I called him in the afternoon. I was lonely, heavy, missing from everything. He picked up, but I could barely hear him over the laughter. The phone cut out, bad signal. He was drinking and on a road trip with friends. I slept the rest of the day.

I took the early train to Aguas Calientes and checked myself into a private double with red peeling paint and a balcony over the Urubamba. The roar of the water through the steep canyon echoed through the empty tin can of my heart. I dumped my backpack on the extra bed, and sifted through my rolled and wrinkled clothes for a fresh sweater. I spent the afternoon wandering the town. It's interesting in that you can only arrive by train, but it's a tourist trap- the end of the road before you reach Machu Picchu. I ambled through the markets, but couldn't find anything I wanted enough to make space for it in my bag. The town is so small that in a little over an hour, I'd walked its entirety. I found a bar with an outside table right by the train tracks and indulged in a few pisco sours, watching other people arrive. *My writing never feels like it's enough to capture the true experience*, I reflected. The interplay of my inner and outer worlds compete for their claim in my perception of reality. It needs a soundtrack. But the only sounds were the laughter of others and the metal brakes on the rails.

The alarm rudely pierced the black stillness at 3:30AM. By 4, I was dressed and packed for the day. I quietly latched the front door of the hostel and walked briskly in the cool darkness to the park bus stop. I was surprised to find blocks of people already in line for the first pick up. Ah well. I joined the end of the queue at the same time as a small guided group and benefited by proximity for the explanations of history and myth. I caught the guide glancing at me during his overview. When we finally boarded the bus, he asked to sit with me. He was beautiful, bronzed skin and sharp features. He told me that he grew up in a village near there. Now he devotes his time to ecotourism projects to help his people maintain their culture. It's an extreme challenge with mining, logging, and the exploits of regular tourism always relentlessly encroaching. He asked the usual questions: where I am from, why am I traveling alone, how long I am staying. He said that when I returned, he would take me in a canoe up the river to visit his village. I smiled and told him not to be surprised when I came back someday ready to go.

The bus stopped, and people spilled out and into the new line at the entry. The guide asked me to wait once I was inside because he needed to tell me something. In the chaos of the crowd I slipped away to start my day, but I was grateful for the brief company. I had a meeting with the mountain and I couldn't wait another minute.

Over the heavily-worn path I climbed. Up, up, up. I crested the hill just as the sun was shattering the horizon, and there it was: the ancient city of Machu Picchu. Nothing about seeing a photo of this place prepares you for how it feels. I stood there frozen in admiration for an unknown amount of time. Time is recorded differently there, and you can sense it in the charge of the place. Machu Picchu remained hidden from the Spanish conquests of the 1500s and was therefore never plundered, destroyed, or masked with Catholicism like so many. Looking down on it, the ruins stood in graceful integrity, their mysteries intact, their splendor sublime.

I found my way to the gate to climb the mountain and presented my ticket. One foot in front of the other, each step a step, gradually I ascended. My heart pounded in my teeth from the exertion and elevation. And then, I was there, at the very top of the world. Over 2000 feet below me were the majestic ruins, now a small smudge in the dramatic scene of giant peaks and canyons. I was sweaty, exhausted, and elated.

An athletic man with a beaming smile arrived just after me and motioned to ask if I wanted a photo. We traded phones to take pics of each other sitting on the cliff. A soccer player from Argentina, he only spoke Spanish and had a difficult accent. He showed me pictures of his team on his phone. They looked professional, and he acted like I should be impressed so I tried to be.

At the bottom of the mountain, my phone rang. It had been days since I'd talked to him. How was he? How was life at home? He was good, thanks. I gushed about the morning's experience. There was an awkwardness between us, stops and starts, formality, too many pauses. Too much distance between here and there. He told me why he hadn't been in touch, and the confirmation stung.

We hung up, and tears streaked down the dust on my face. I sat at the edge of a tall terrace and let my sneakers dangle off the side, kicking childlike against the wall. I couldn't blame him, and no rules were broken, but I wished that somehow things were different between us. I wasn't there, and even when I *was*, I wasn't really. God, we were both so lonely.

The soccer player walked over and plopped down, happy to find me again at the bottom of the hike. I started to explain that I wasn't in the mood for company, but his smile vanished when he noticed I'd been crying. He turned my face to him, smudged the wetness on my cheeks with both hands and kissed me.

When I wore the shackles of control
All I could think of was freedom
and how they chafed against my skin,
How I'd explode into creative authenticity
if only I could pry apart my bondage
How I'd run and play and revel in endless wildness
If only I could free myself from the trap you set for me,
The cage you made for me.

Now in the vastness of choices
And endlessness of options
I am a rudderless ship tossing in the foam
of an overwhelmed sea
Where is that exact self that begged for expression
when she had hands on her throat quieting her,
Shushing her into obscurity?
Where is that girl who knew precisely what she wanted?

What if a damaged me
Needs the damaged you
For clarity
For contrast
To see and finally defend my Self against someone who
hates her more than me
The idea sends me reeling and I
force my self to the surface this way.
Never again
Never never never again.

I don't need my freedoms to be revolutions
And I promise myself to quietly love harder
that detached part of me
That struggles to acknowledge the validity
In any choice that I make,
In any path I wander,
For whatever reasons I have, whatever impulse I succumb to.
The judgment and insecurity that paralyzes me,
That walls me off from decisiveness,
Is an echo of your voice inside my head.

Ruins

The places you told me about
Are places you remain
Even after you've gone from them,
Gone from me.
Your touch remains *here*,
And here,
And I hear your voice now like a prophecy
For a self I wouldn't yet know,
A trip I wouldn't yet plan.
I am alone in the image you evoked so many lives ago.
Echoes of an Us
As ruined as the crumbling fortresses.

Dragging my fingers on the walls in a dream
Day drunk and quite glad you aren't here.

I think traveling is preparing me for the day I lose my mind.

I move through countries whose languages I don't speak well or don't understand at all. You get used to not really knowing what the hell is going on, what specifically it is you're eating, what time things are "supposed" to happen. You start to relax into doing what it seems other people are doing, therefore you subconsciously notice and imitate the small things about their manners, their choices. The sooner you can quiet your mind and become as local as possible, the more comfortably you can move about a place and access its deeper layers.

At first, you learn this dance by simplifying. You develop a certain primal gratitude for being able to find a safe place to sleep, a way to get money, exchanging money for food; you have victories of feeling like you can provide for yourself. As you integrate, your victories become more nuanced: trying new things, finding things you like, adding phrases in the native tongue. Typically you can rely on cues from other people as to whether there should be concern or calm, decorum or severity. People line up, I get in line. People go, I go. I have to pay attention, but at the same time, cede control completely. I can get my basic needs met by mimicking the behaviors of others until I understand more in a place.

Things start to happen when they happen. Or rather, they keep happening when they happen just as they always have, but *you* stop concerning yourself with exactness. It's not when the bus comes but if. And you develop little back-up plans to keep yourself a step ahead, but you go back to relaxing because you have all you need with you: an intuitive and adaptable self and a bag full of comforts. You notice yourself adopting the pace of a place: slower and more peaceful in nature or villages, quick-thinking and adept in city traffic. Traveling forces you to pay attention, and there is nothing greater than living your life in the discrete minutia of the present.

You can count on yourself to survive without overthinking

everything. You'd never expect this in the culture of control we adhere to. So many of us rely on schedules and devices and routines and the absoluteness of our expectations to find any comfort or functionality in the world. The truth is, these structures have caused us to lose touch with instincts that connect us to our surrounds more deeply. Traveling makes you remember these, awaken them. You can check in with yourself in any situation and peel things back to survival- and almost always, you still have everything you need to continue living. Being uncomfortable is ok. In fact, sitting with discomfort is catalyzing for the growth of the Self.

Maybe if the day comes that I degenerate into more severe mental dysfunction and I lose my reality, I will be able to use that same learned calm that I've gained and practiced from my travels. Maybe it will be instinctual to move through unfamiliar situations with the comfort of my self-reliance, trust in fellow humans, and a general sense of peace that everything will work out.

The sun drops from the edge of
this place, absolute.
Another day slips through crimson fingers that
can't bend to grip the rope or cut it.

I slide into the draped silks of the abyss with all the rest
And for once I don't miss you at all.

I don't fight the relentless movement of the clock anymore,
the irrational way it paces itself:
Steady and removed and dissonant
against the erratic backdrop of my internalities,
The moods of time within me
laced with pause and sprint.

Let me slip now into the hours of water and weight,
I'll turn away with the spinning coin,
Let me temper my universe to match the little deaths
Turn, turn, turn away
until I face the death of you
And your confinement to a tomb of pages
That I can't read in the dark.

Some people need a lot of control in their lives to feel ok. When I sense that side of myself manifesting, I make war with it, force my monsters into the light. Any time I feel that pull of needing more structure or dominion, I question it, argue it, expose what it is that limits me from having freedom in all facets of my life. Free living doesn't mean there is no method, the lack of rigidity doesn't imply lack of form. Of course there are balances to be achieved to be at peace. But the secret is in the openness to adapt, the eagerness to play. I've been hanging on too tightly for too long, and I'm over it.

We tend to create little obstacles between us and freedom. How can you tell what's holding you back? Think about the judgments you cast constantly to screen what's acceptable in your life, think about the things that make you upset that have nothing to do with you: other people's choices, the weather, expectations of events, the illusion of permanence. Then ask yourself *why* any of this bothers you at all. Any time we get bound up by trivialities, things we can't change, brittleness in our thinking, or analyses of others, we miss an opportunity to be free.

I'm traveling with an older demographic on a tour bus and I notice the difference in the severity of need for comfort, both physically and within ideas. Is this something that naturally occurs with age? Regardless, how can I fight to not only retain my adaptability but also to expand it? Conversations tend toward judgment ("He is eventually going to have to get a real job", "I could never sleep in a hostel"); or the infinitely boring labeling of things as one pulls them into their mind, their story, and therefore takes ownership of them ("Even the police car is broken down", "There's another llama", "Ooh, lightning"). There is no profundity or extended thought process, there is not even real surprise, excitement, or concern. They are not living, they are narrating to keep their fears at bay. They continue to prattle on about nothing because silence is too uncomfortable. The semblance of control is easier to maintain with constant self-reassuring vacant chatter that goes unchallenged and therefore creates the illusion of a shared experience. I'm in my head again today, the anthropologist. No one offers anything authentic so I just smile and return to my thoughts. I don't have the energy to change the tone. I'm surrounded by people but incredibly lonely. *Water, water, everywhere...*

Today's crowd on the long excursion to Puno contrasts sharply with the lovely company I've stumbled upon in less canned transportation. My mind drifts wistfully to the friends I made on the train to and from Aguas, connections that I hope to will somehow endure. Today there is nothing but small talk, which makes me recoil into introversion.

Ironically, what tests my comfort zone the most is spending time with those who neurotically depend on structure, on schedules, and on the fixed narratives they've crafted. *See how I'm judging them? See how I'm missing out on my own happiness in the moment?* Shallow interactions burden me with my own residual similarities, and trigger me into reflection on the work I have yet to do to liberate myself. I don't want to spend my life asleep, driven by fear, confined by my own delusion of safety in mundanity. These are things I increasingly reject, correct within myself. I want messy and real. I want to know the truth of this place, I want to feel it. The plush seats and broad glass windows are a cage that keeps me insulated from a world I long to touch as we barrel past it. But this barrier preserves the separation that the majority of this group seems to depend upon in order to participate in a culture so different from theirs: they want an observation rather than an immersion. This makes absolute sense, as witnessing begets judgment from its stratified nature, whereas experiencing inspires empathy from a shared perspective. Recognizing the irony in my own self-seclusion on the bus, I almost laugh out loud. Maybe I could try a little harder to participate in *what is* instead of wishing it away. There's an hour left before we arrive in Puno, enough time to at least try to have a decent conversation with someone.

Time moves oddly and folds back on itself like one of those quadrangular origami fortune teller games. Stories unravel between the lines. There are a thousand ways to tell the truth, depending on what you focus on.

Last night I stayed with a family on Kantuta, a floating reed island in Lake Titicaca. I was the only visitor. In the evening, I set nets for *chalua* from a reed-and-plastic-bottle boat with the family's patriarch, Eduardo. The lake was serene with the daytime tourists gone. Music drifted over with the wind, and the lights of Puno shown like glistening embers on the hill, but we were separate from the bustle and grime of the city. Communication was peeled back to showing and feeling, as Spanish was a limited second language for all of us. It's always so magical to me how much comprehension goes beyond what is merely spoken. Eduardo and his wife Maria taught me some words in Quechua, how to make medicine for headaches with the reeds, how the islands are constructed from the roots, and how to gut a fish. Maria and I fried the little piranha-like fish and boiled rice for dinner as the thunderheads started to build. The clouds let loose as we were finishing cooking. Laughing, she stuffed my plate into my hands and motioned for me to run my hut as the family ran to theirs. There wasn't enough space for all of us in one. I jogged across the springy, wet reed mat and barged through the small entry into my space. After changing my soaked layers and adding some fingerless wool gloves, I sat in the open door and picked the tiny bones from the fried fish, watching the sheets of rain drench the lake and land. It didn't let up until morning, and all too soon, I was on a speedboat waving goodbye to my island family. In the early light, the boat wove deftly through waterways cut into in the reeds, lake roads with no signs. We pulled up to shore where two men in a randomly parked taxi met me on the embankment.

Today, I'm on a 3 dollar bus winding through the hillside, sitting next to a lady in traditional clothing. Somewhere in the back, chickens squawk. My seatmate knits while I write, but the ride is so swervy that eventually we both give up and look out the window instead. Over mountains and through small villages we wind, at speeds that challenge the coach's center of gravity. I'm content even though I'm hot, uncomfortable, and half-sick from the ride. The scenery is remarkable. Mud bricks drying along a hillside. Heaps of dark red corn drying on tin rooftops. Flamingos wading in high altitude lakes. Vicuñas on an arid plateau.

Inside, the landscape mirrors this strange external world. I marvel at what I'm letting go of, what I miss, and my discomfort as the exact place from which new growth is possible. The old pain is forceful, but the medicine is subtle. When I can get quiet enough, I can feel it working, softening me.

The sun pierces the gray din of clouds just as it's setting and we are cresting the last ridge on the way into the volcano-lined town. In this moment, I am at peace.

I'm at a bus stop in Lima, and I'm edgy from getting ripped off on the taxi ride here. And I don't understand where I'm supposed to be and when, or where my backpack ended up. So I have to stop and remind myself that nothing matters.

I was in a crumbling hostel in Arequipa last night with no water and then ice water, and the clothes I sent to the laundry got lost. And nothing matters.

I remind myself this like a mantra, I breathe into it: I have myself. Water, food, notebook. I could be anywhere. I'm complete. But the anxiety can sneak in on these travel days because they are Days of Expectations. I expect to get on the right bus at the right time, I expect my bag will go to the same location as me. There is a time to arrive, a place to be.

If I can't let go otherwise, then making so many plans and inviting these expectations to roost is how I fail myself.

One day soon I need to travel with fewer plans, no plans. Then I can test whether I actually feel more free, or if the lack of defined anticipation would make me even more unhinged. *I'm going to teach myself freedom.* I cannot become one of those rigid old people on the tour bus: fixed mindset, terrified of change, death, or surprises. Let me run toward all that I fear not because I'm brave, but because I'm tired of being conditioned to be afraid.

I'm going to pry my clenched hands off of my life one way or another. Intention, methodical therapies, or trial by fire. I don't want control. There is nothing to learn from a regimented, predictable life.

You are potent,
fecund with an infusion of the unexplored.
Ideas disrobe for the thinking,
Thoughts blossom into conversations and
Drop their seeds on the earthen floor
for unpromised tomorrows.
I have learned by now to leave them
unswept.

I love when the radio matches the
soundtrack of my dreams:
A deeper remembering,
A tug on the bell chain at the portal of consciousness
snapping me back to awareness of
my hands on the rough-hewn doorframe.
I exhale gratitude unilaterally,
Clean of constriction,
As a universe conspires openly for
A life on fire.

I waited for a time
because I've never been patient and
because there is a difference in
that which bends the boughs
In bored abundance and
That which is rare and cherished,
glorified by the careful eyes of the long-starving.

I needed to feel the quality of my emptiness,
to be at peace with the sharpness of hunger
And learn to guard my fingers on the take.
The pause sings now with intention in my cells
as the most delicious of moments,
Abjuring abandon only for that which is ripe and real.

Nourish me now with the stories of your arms
They've spent their time learning the sea
While mine ache from
opening the throats of mountains and

clawing at the citadels of fools.

Tell me the contrasts you've known,
the moods of the ocean,
Tell me what keeps you sleepless
And what makes you free;
I'll liberate the myths from my scars
as locked eyes unknot our revelations,
Indulgences in connection...
The juncture now at which our narratives collide.
Tell me you savor this as I do.

Meet me here without the masks of flesh and translation
I have learned to transcend these things without poison
Give me something new,
Something deprived of comfort.
Give me the upheaval of coming unchained
from all the things I trust
By leading me past the uncharted periphery
of all that you do.

I want a passionate life and
if I can't find it in one place
I will collect the ceramic pieces of it from Everywhere,
One by one,
And glue them into a cup to hold my vast and waiting heart.

I long to overflow my brim with the
waters of wild rivers and the
fiery liquors of lovers…
Tiptoeing along the brink of an ever-changing edge
that I will stop at nothing to uncover and tempt
so that I can feel the limits of this container,
its exactness in here and now and
its expansiveness into inner space.

Flesh and emotion boiling with reality and life
Kinetic and challenged into clarity
Learning the infinite by assimilation of all that is not;
Finding the dirty face of god and kissing her,
fermenting her
Crushed grapes in bare hands
distilled into brandy to be spilled indulgently in
crashed cup dinner party cheers-
and also sipped so delicately,
aromas and nuance inhaled in lingering delight.

The flush of both dimensions washes over me
and I know the wholeness of my soul from contrast,
Space and negative space,
I know a home in any place I wander
and no home whatsoever,
Hung between the cruel elucidation of mortality
and a manic euphoric dream
that allows simultaneous universes to exist within me.

From the discomfort and beauty of both I fill my cup
And effervesce with gratitude for finding liquid abundance
after escaping an arid life
in a closed and empty bottle.

I have lost the need for a solid state of self, and I am moving through the currents of life, of place and mood and circumstance, entrained and floating, exonerated from carrying myself. I surrender. I deny myself the urge to fall into obsession with if, where, and under what circumstances I will land.

What the mind becomes engrossed with can be a reflection of what is lacking within, I'm sure... so I study the things that have a gravitational pull over my heart, the things that would anchor me into fixation if the river slows or I succumb to the weight of my neuroses.

What am I running from or to?

I want love, insanity. I look for it, invent it delusionally, numb myself from craving it. I want to feel reciprocal fascination, to be mutually dissolved in the liquid of the other. I have an incredible partner, but it's so... complicated. Wounded now. His lack of *need* of me, desperation for me, translates often as passivity bordering indifference. I am not sure he really *knows* me, but then again, I am not sure I know me anymore either. I message friends; I pore through my phone as though the answer is in receiving token outpourings of attention... as though some number of pieces could be glued together into a whole love. The need is insatiable because I'm a leaking vessel with no way to retain what little love I can offer to myself.

This is the sickness of my psyche, I see it now but can't change it yet. Only with obsessive love has my achingly parched heart felt quenched, only when someone is truly mad about me. It's dangerous to think that this is the game I play, subconsciously drawing this to me. Surely it's a withdrawal reaction from the beautiful poison of a crazy love that almost did me in. I was addicted and now I'm clean, but my heart just wants a new drug. It will take more time to heal this than I thought. So here I am, halfway across the world, in suspension.

I no longer know how to feel or accept a tame, grounded love. And it's impossible to keep an explosive, feverish love, so I'm condemned to this anguish and loneliness until I am rehabilitated. I am unfed and hungry. Or fed and yet unnourished. My heart twists in angst.

Acceptance of this ache is a new development for me, an idea just awakened. I've been focusing so hard on trying to hurry up and heal, but it's like I've been in shock, never really acknowledging how badly I'm hurt. The seeds of a new understanding of myself have been sewn into my consciousness, and I water them by sitting with this, gently. Healing comes packaged in these small increments where it stops hurting long enough to actually look at the wound.

What if it's ok to just ache? What if I could accept that I'm a Lover, but I could manifest that love myself instead of depending on someone else to mirror it back to me?

How can I become the thing that I need the most? How can I find comfort in this mutable state, stuck not in the longing of search nor trapped against the rocks on the bottom, but flowing ecstatically as part of the river?

Will I ever really be able to go *Home*?

I called my Dad, and he asked how I'm enjoying the place I'm in. I offered the truth: I find it lacking any unique identity. It's shallow, tourist-infested, and despite its stunning photogenic quality, Huacachina fails to enrapture in real life. I didn't mention that every time I leave the hostel, I'm catcalled and hassled by the same small group of local guys, each time with more overt creepiness. The town is so microscopic that they are unavoidable- at this point I know that I won't even be going to get dinner tonight. No doubt this and the inhumane heat have added to my disenchanted impression of this oasis paradise.

He seemed surprised and remarked that it was unlike me to travel someplace I'd review with negativity or indifference. This irritated me, as it's categorically untrue… and plus I'm feeling impossibly lonely, sensitive, and physically and energetically trapped here. I called because I needed connection, and instead I found further frustration. I rarely complain about the places I don't enjoy, but to say I don't experience them is grossly mistaken. I momentarily tried to make him understand because I needed to feel seen, but I quickly resorted to downplaying my feelings because the conversation was too complicated for the poor connection both literally and emotionally.

Of *course* I go to places I dislike. It's part of the unspoken accord to which you agree as a traveler as opposed to being a vacationer. You will see much, but you will pay a price, and it's never the price you think. You pay it in boredom, wasted time, thwarted expectations, discomfort. Sometimes in actual danger. But in those very things is opportunity for true growth and the illumination of scale and meaning. This contrast is the only access. I never regret seeing the disfigured, the overt, the crass, the poverty-stricken, the overprocessed, extorted, false, or superficial. These places are Real. If they have to exist, then I want to know of them too. They give the others richness, and you *have* to see them, feel them, to understand. I would never expect or even want to like every place I go, every person I meet, every song I hear-there is a drab bypassing in that which seems to go against an internal truth. I see a lot of ugly places and I go through a lot of ugly things to arrive at beautiful places. I have found beauty where I'd have never expected it. I have seen the pain-encrusted underbelly of the most remarkable wonders of the world. The way I experience things is multifaceted, honest, complicated. Maybe we experience things more as *we* are than as they are.

I spent the afternoon by the pool, existing, and not much more. I was overstimulated, overexposed, and exhausted. The thought of the man grabbing my arm in the street earlier rolled around in my head. I felt so far away from home.

How could I blame anyone else for not understanding me when I barely understand myself? They don't ask any *real* questions so I don't offer, but this time I caught myself going through the motions of insulating someone with an illusion. The widening of the gap between my image and my reality guts me, and I'm empty and misunderstood and numb. Invisible.

I plastered my social media with the obligatory pictures of adventures and smiles, meanwhile in my room I turned the air conditioner all the way down and pulled the covers up over my face, waiting for sleep or death or tomorrow with relatively equal indifference.

It just came to my attention that I no longer suffer from the effects of the crippling sensory overload and dissociation that impacted me so often when I was with P. My anxiety is still present, but it is changing. Before, there were so many times that I became overwhelmed because of noise, or contrasting inputs, or gaslighting, that it was like I was literally losing my grip on reality. I would sink into my internal world, retracting into myself as a form of protection, spacing out in an unbearably dysregulated nervous response. I knew back then that this was happening with increasing frequency and wasn't normal, but I had assumed that it was something worsening with age, an inevitable genetic or degenerative ailment. It never crossed my mind that it was a "freeze" survival mode, a disconnection with my body to endure a situation.

Months after the trauma of all that was life with P, and then all that is life without him, I have noticed out of the blue that the panic attacks and dissociation are gone. I've been in circumstances that would have easily triggered these in the past, and I now come through unscathed. Or at least- I am able to stay in my body under stress.

I'm only just uncovering the ways that I became damaged in that relationship and its aftermath. I find these things out in moments when I realize I'm *ok* now, and only then can I see how badly I wasn't before.

So sad I still am for the child within me that clung with both small hands to the burning house around her.

The more I see of the world the more I fall in love with it. Not the glittery smooth skin of infatuation, but *real love*, love for the bones of places that may be healed and calcified into deformity from breaking.

Or not healed at all.

Love that sees the peeling sunburned face, the taped toes, the hunger and hindrance and humanness. There is a shared breath between us in all our forms and facets. There is a shared heart between us and the Mother that sustains us.

There is a will to *live* and *take* that grips us like the throes of fever in mad dog desperation against the struggle. There is also a will to nurture that catches us like a golden net. It unfurls just before the free fall ends in the personal tragedy of hopelessness, or the collective tragedy of the deterioration of humanity.

For those of us with a certain spark, there is a will toward leveraging our insatiable curiosity about the manifestations of the human spirit into healing; a will to bridge the disconnect and ground others when the panicked, rabid nature overwhelms them or Us... a will to touch the wounds and know the gaps so that we can bridge them.

Once you start thinking like this, everything just looks different. Though I'm still adjusting the focus of the lens, everywhere I look, I see a chance to love harder.

Ausangate loomed in the immediate background, a flowing scarf of stretched fog catching on her sinister neck as the winds and sun vied for domain. Light dappled the spongey moss-covered ground in a kaleidoscope play of grays upon the chartreuse vibrance.

It was surreal.

The scenery was hung in the paradox of the extreme brutality and extreme beauty such places hold simultaneously.

I stopped to drink it in, and Alex, the guide caught up to me. He turned to admire the giant. "We still worship it, the mountain. Even though Catholicism is the major religion, most of us believe in the Old Religion too. My Grandmother never converted and continues to worship as the ancient Incans did."

We turned to divert our attention to our footing as we ambled the path back down the mountain. We'd just summited at over 16,000 feet for a view of Vinicunca, the Rainbow Mountain. It was a challenging peak to reach with the impacts of elevation, but was dwarfed in the footprint of one of the most prodigious giants of Peru.

As we walked, Alex talked about religion, the arrival of Pizarro and the Spanish influence, and how Christianity has never completely erased the ancient practices of any culture, as much as they tried to stamp them out with the overprinting. "In the Cathedral in Cusco, there is a painting that shows The Virgin Mother, Maria, as a mountain. Most of us still do rituals for the gods of the sun, moon, water, the mountains… we are tied to the earth and to our ancestors."

He told me about his family, his heritage, his life and their lives. And in those moments it dawned on me: in the US, in our assembled culture, we are descendants of many races. Most of us have little idea about our lineage, our ancestry, our narratives- the stories that tie us to our ancients. We have no long-standing connection to our places. We have lost the thread that binds us to the fabric of something greater than ourselves, and the effects of this are personally and collectively devastating. We are orphaned from our roots.

This is why I travel. This is among the reasons why I want to know the deepest stories of other cultures. Further out on the spiral is an even greater look back at all of us: a story in which I, too, am a part. If I can't hear the stories of my own family, my own ancestors, because they are lost or not offered, then I will listen to as many of the stories from the others that I can. It's in the assimilation of these that I gain the cosmic perspective of inclusion and can reattach myself to the feeling of being within this human tribe. It's in the arms of the strangest strangers that I am reawakening my ability to Belong.

LA

Before the breaking I was being bent.

And before the bending I could feel the energetic tension vibrating around me, pulling me, weakening me. I could hear the tinny high-pitched singing of the tight rope, the swan song of the stretched-thin self divided between two worlds.

It's never as much of a surprise in hindsight as it feels the moment that you shatter. Maybe we call to us the circumstances that will catapult us into the growth our souls need to face, so we subconsciously recognize the consequence we will come to bear. In reality, hundreds of small choices can each have integrity, while our blind spots keep us from seeing that we are breaking beneath the weight of them together.

The point of surrender is shocking when it trips, we overtrust our own malleability when things pile on too quickly. Pieces of ourselves recoil explosively, then dangle in atrophied disconnection into the expanse they once reached across.

When a dam breaks, the river forgets the ache of its suppression and knows only the free fall into banks familiar yet foreign, a bed where it lay in a dream, a channel carved by some ancestor-self to show the way. But although the river knows only the present moment, its character is not discrete and spatial- it is differential, adaptable. Its expression is not defined by its past, but by its relationship: it has shaped the land, and in turn, the land shapes it. Its identity is complex and lies in an evolution, much like our own. We are water, moving. We are capable of only reconstructive, forensic remembering: we have changed the landscape over and over again, sculpting everything around us with invisible and mutable hands, but we cannot see our ever-changing selves or our current impacts clearly. We become impeded either by our actions or by no fault of our own, we gestate in the pause, we find release and continue on.

We are similar to a dammed river blowing out in our rupture into awareness, in the birth that delivers us into a world beyond complacency. The breakthrough can be violent, but the experience itself is aliveness on the cusp of divinity. It can feel like complete destruction, but on a deeper level, we very much know how to do this. Our freedom from previous limitations still only has the energy of our potentials. It may feel like we will fall forever, but we will find a landing with boundaries that we ourselves have determined. The negative aspects of such initiations exist largely as constructs we create in our anticipation of consequence or our fear of change. Upheaval is a chance for communion with our impermanent and transcendent nature, but it is not easy to cede control and allow this seed of thought to be born into cognition.

In our adult consciousness, there is choice. We can view the trauma of our awakening as terrifying and abrupt, a forceful and unwanted delivery from safety and warmth into a cold, scary new way. Or we can witness this event in presence, we can see all parts of it, even the pain, as the very essence of our humanness. We can find curiosity in the unknown, exhilaration in bursting forth, vitality in the strangeness. All our perceptions of ourselves and our reference points in time and space are obliterated internally, a letting go. In our breaking is the possibility to understand the pure nature of Self, but it requires us to submit to openness in the midst of our most difficult trials. We have to stop the narration that looks backward in replay or forward in expectation and allow ourselves to be carried forth.

It's instinctual for us to fear change, it's a remnant of our wildness, it's self-preserving. We are also wired to ruminate on our past much more easily than we can imagine the future. Our biology can work against us if we don't lift ourselves into higher consciousness. Our memories are only emotional abstractions within our narratives, as pliable and mercurial as our self-image. They change across time, as our perspectives change, and they only impact our *perception* of ourselves, never touching our real selves. A river doesn't romanticize its days of being rain or a being lake, it doesn't lament the way its banks have eroded or how it once flooded. A river just *is*. If we can learn how to just *be*, we can fully accept each moment of life, whether we are in flow or cresting the brink of massive transition.

Our circumstance is our accountability.

What do we pile up in our narratives to appease our egos, and what is the gap between that and what is Real?

It often takes a leveraging event to snap us into objectivity and see ourselves outside of our stories and relational references. This is the point past which a deeper understanding and peace is possible. Breaking seems to be the key to accessing this wisdom- there is a barrier between our superficial selves and our true selves that is not transparent- it must be destroyed for us to achieve this degree of reflection. Is it ever possible to orchestrate these episodes? Breaking is, for most of us, external, unwanted, and scary when it happens. But if we can let go in this process when it occurs, it becomes the catalyst to our highest transformations.

There are truths about the self in layers. You are your body, but you are not only your body- you occupy it and utilize its sensory receptivity to learn and feel. You are here and now, but because you are capable of directing your consciousness, you are not *only* here and now. Where is your mind?

Our limitations are dictated by our biochemical machine unless we train ourselves to become more subtle, wedging our minds into the cracks of a shattered illusion to keep it from closing. Some people may only ever have the body self, the Animal- and the finite understanding that is born from an existence completely within the material reality. But there is a selfhood so much more vast if we can escape this plane to experience it.

The question mark hangs itself in the silence,
The wall between the wondering
And the knowing
A wall with windows but no doors

The body is a shell you get to borrow for 80 years and then return. *Unlearning attachment* means non-attachment to your person as well. When you can internalize this, you respect your vessel and take care of it so that it can carry you through your years, but you no longer fear the signs of aging, you no longer fear death: you are free to explore the outer and inner worlds to the extent your body permits, with a full understanding of the contract your soul made to come here.

We spend endless efforts and energy trying to re-create. We expend infinite mental energy mired in impossible longings: wishing to reignite an old love, re-inspire an old feeling, revisit an old place. But what else *is there*? What if we were to dive into the unknown, unfolding present with the same fervor with which we indulge in our nostalgias?

Once I get my heart buttoned back up and these tangles unknotted, I will spend my time Free, swimming in the wellspring of life emergent.

I unblock you sometimes
To look at my old life as an
Outsider.
It's bizarre how separate
You can feel from something
In which you were once dissolved completely.

Every time I'm able to look at you from here
I let that life go
a little more.
I am grateful for the peace I have now
that I never felt
from inside there.

This, here, is the world where I belong.
I breathe in through the open window
With no one calling to wish me away from this moment
At home.

We are animals with front facing eyes. Predators.

Manicured desert city strip malls
With their perfect succulents
And mist-machine sun prisms-
All of this reminds me of you.
Happy emptiness, illusion.

Palm trees
Keeping their secret loneliness in the drying fronds
The ones that no one can keep up with trimming,
The ones underneath,
furthest from the new growth.

There is pain hiding under the surface
Of something made to look beautiful.
There is the realness that no one wants to see.
There isn't enough water here
To stay alive for long.
I died in a place very much like this once
And a little bit again every time I come back.

Now these artificial oases are
a garish and unnatural headstone
perfectly fitting for the me that loved you.

Cuba

I don't want to *vacation* in a place, I want to embody it, have a relationship with it. I want to see things for myself, I want to *feel* my way through places I would otherwise only know from media or books or what someone else has told me. I want to touch the silk and leaves and sandpaper and crumbling concrete of this textured physical and emotional world... I want to develop my personal range of senses to appreciate this life in every capacity, learning and indulging in all the varieties of ways that humans can present outlets for experience- in cultures exotic and ripe for the tasting.

I travel so that I can expand the range of my pallet and savor every moment of an existence that is given vibrance by exploring these contrasts.

Havana, in that way, does not disappoint.

I read what I've written previously in my journals with objective curiosity. I'm discovering a self that I no longer am and no longer know. In fact, I'm discovering many of these selves, each separate and stranger than the last.

There is familiarity, of course. I'm not removed from the person who lived these things and felt these feelings, I've just layered so many new experiences on top of them by now that there's distance. I've changed so much even in just a short time. Maybe I'm healing. The poems trigger scenes I recall like films I watched as a kid or dreams I had years ago. I recognize my voice telling the stories, but if I hadn't written them down as they were happening, I know I would remember them differently now.

Every now and then, there is a strange consolidation of all these disjointed selves, as though the entirety of all I've known and lived through is internalized at once. My very existence vibrates with some kind of *knowing*. It translates within my body simultaneously as the ache of longing, loss, and infinite love. It's as though my cells can remember what it feels like to be molded and remolded, kneaded through the extents of emotional ductility, structure, fracture, and resilience. The memories are experienced bodily instead of mentally. In these moments, my Being can see through the layers of things that cut and impress upon us, the themes within the collective consciousness that bind us all and make us human. This feeling validates and commemorates all my former selves that made all these choices, even the mistakes, because I was doing the best I could at that time. It's self-compassion, humanization, forgiveness. And yet, for the individual stories there is still always the sense of being removed, of poring over someone else's thoughts. Maybe there are still some things I struggle to look at squarely in the light of day. There are selves yet slinking behind the curtain, waiting in some ashamed unbelonging for their turn to assimilate into the wholeness.

For all the different selves I have been, I write them and then read them to truly know them. There is no automatic understanding of my current inner space without undertaking the process of creation and subsequent observation. I haven't always been this way, but this is me now, actively healing, pulling all this back together. It's abstract and a bit uncoupled to have to write to understand myself, but there seems to be no reconciliation of my fractured psyche without the process. If I go too long without writing I just lose myself, these selves, all the more. Introspections that I fail to commit to paper get consumed and digested within looming larger thoughts. Sometimes for this reason, my writing is frantic. Maybe I am afraid that if I don't capture my thoughts at these pivotal moments, my memory will fail and I'll forget entire selves I have been. I have a need to collect all of them and reflect on them without the bias of my perpetual metamorphosis reconstituting my memories to suit my current perspective.

Is the desire for this methodical self-analysis a result of processing the recent loss of my identity? Or does it stem from the bigger ongoing questions of selfhood?

The journals and poems are my portals back into the emotional realities of my past, snapshots of who I have been. Without invoking that context, there is no meaning to be gleaned, no way to know how I truly felt as I lived through something. Without tapping into this, I run the risk of becoming delusional about the things that have shaped me. Putting myself back together right now requires the complete collection of selves to be assembled and studied for clues. I am lost, but I have a map. I have a list of waypoints and intersections I passed, the way that I arrived here.

I am in constant evolution, a process that I feel happening moment by moment, and faster than ever. Even if I can't pin down a "self" until she is written, change is something I feel with every expounded thought. Maybe this is why it seems like reading a stranger's journal. To write one's deepest thoughts is to own them entirely, to take power over the direction of the narrative... to distill it from the holistic and commit to it linearly. I file these things in history and can mull over them later, freeing me to move forward. Pages and pages of selves, slices of time and space, perspectives that can be accessed externally if the words succeed in capturing my qualia. They make it accessible for my future selves to retrieve unadulterated renditions of my past.

By understanding exactly what we have undergone, we can reconcile ourselves and correct our erroneous lines of thought. We can find ourselves and understand our context. We can see our reactions clearly and begin to scale them more appropriately. We can rectify our sense of place and purpose should we ever make the mistake of allowing them to become diluted.

Right now, this seems like an important measure, a means to have an accurate picture of myself. I don't write from the top of the mountain, looking out across retrospective lucidity, as though some prescient awareness has led me through this transformation. I write from the thick of things, my entries from the trenches. It's real time, a totally different tale than any that I'll surely have later upon looking back. I have decided that it is far more important to *become* great and wise than it is to be *perceived* as great and wise, so these are stories of Becoming. It's raw and complicated and vulnerable to Become, it's an undoing. The immense pressure is crystallizing the diamond in the mantle. The seed is bursting apart in the warmth of the sun. The wet wings are emerging from the chrysalis. The old is being destroyed and transfigured in ways that can feel impossibly devastating but that are surely are natural and necessary, wiping the slate clean for new beginnings to arise.

A broken window crank rattled in the dirty dashboard tray as we bounced along the Havana Street. *The taxi driver must use the single crank to adjust all four windows,* I mused, noticing the star-shaped stobs that stuck out from the faded vinyl door interiors. The windows were all down and yet we were sweating relentlessly, sticking to the worn red couch seat in the back. The heat was oppressive.

Callejon de Hamel. It was nothing more than a name I'd written down from a forgotten guide book or maybe a travel blog I'd read before I left. I no longer remembered what it was or why I wanted to go there, but we had free time after going to get bus tickets to Trinidad, and Bex was game to check it out.

The brakes protested as the old Chevy came to a stop, and we poured from the car into the mouth of a narrow street. The scene was pulsing with the enchantment of percussion funneling from deeper within the canyon of brightly painted multi-story walls. I gave Bex a shrug to check her interest; I had no idea what we might be in for.

The lure of rumba music led us past metal sculptures and bathtubs repurposed as lounges. In the mix of sun and crowd we wandered into the heart of the party, which through the sharp-hollow hammering of drums had taken on a life of its own. The rumba consumed those who entered: a mass of sweaty, heaving, happy bodies surrendering agency to the throes of the music and rhythmic chanting. We worked our way slowly through to the other side, suspended in indulgent fascination and sensory overload from drums, people, heat. We reached a break in the crowd when I saw it. An altar, dark and magical, resplendently adorned with bells, symbols, and a black-painted figure. I paused as my memory recovered the reason I had wanted to come here.

"This is the *real* local party," a man behind me said brightly. I turned and met a smile that matched the tone. He introduced himself as Mechel, and went on, "The altar is Las Reglas de

Palo Monte."

He ushered us through a small doorway as we talked and I told him that the couple who had presided over my wedding, some of my best friends, practiced Yoruba.

"You'll appreciate these things then," he said in a mix of Spanish and English, leading us down a small set of stairs and through a veve-adorned black door into a concrete room. The room was refreshingly cool with air, heavy with content. A few other people sat in chairs or stood talking, and there was the impression that this was a space one had to be invited into. Brightly colored paintings and sculptures lined the room, their vibrance both accenting and contrasting religious regalia and references for a culture I knew only a little about.

"This whole place is the work of Salvador Gonzolez Escalona. Come, since you know of Yoruba, I'll give you a private tour of his house." We followed him across the alley and back through the flowing mix of faces: all of them high from the ritual and rumba; laughing, sweating, enthralled in the excitement and insanity. We crossed into an open doorway of a bar, dimly lit and private, occupied again only by the invited and initiated.

"Look," Mechel pointed at a wall of photos. "Some famous people have connections here that you might not expect." He laughed. I recognized some of them as actors and public figures from the US. We walked down a hallway and paid respect to an altar that had expanded into a whole room. Conch shells, candles, coins, and bones were piled around the central platform, and brightly colored fabrics draped from the ceiling of the enshrined space. I stood in awe of the assemblage, said a silent thank you in gratitude, and added some coins to the rest.

We entered a shaded back courtyard and were greeted by others- some local, some tourists, all there by their own winding stories and a litany of unraveled circumstances.

Soon we left the casa and made our way dancing back out and through the hoard of a hundred or so pulsing bodies. Some people were clearly day drunk and partying, others in a state of trance from the drums. Mechel wandered with us back toward the entrance of the alley and into another gallery, then offered us CDs of rumba music. Having no disc player, I declined. I was getting a bit worried that Bex was overwhelmed by the crowd and heat (as admittedly I was), so I began to excuse us from the party. I gave Mechel ten CUCs for his time and in support of this fascinating underground community.

He leaned in. "You really need to see the Babalawo before you go. You believe in such things. The Babalawo can divine for you. I can take you to him if you like."

Refreshed by our interest in this tangent adventure, we left the callejon and walked together a few blocks, past the crumbling buildings and littered streets of deep Vedado, Havana.

Our destination was a dilapidated three-story complex, indistinguishable from others that had been abandoned. The small door was open a crack and lined on top by a garland of dried palm leaves. Sensing my gaze, Mechel explained, "These are marked houses. It's how you know where to come for Palo or Santeria."

Through the entry was a basic apartment. Simple. No decor. We followed him up a stairwell that lined the right of the bare living room area. On the second floor, we found the Babalawo, an unassuming man perhaps in his forties. His skin was a burnished bronze that evoked the complex lineage of mixing races in Cuba. His plain clothing and dignified presence gave nothing away- there was no indication, save perhaps the intricate beaded necklaces slung crossbody, that gave him distinction as a revered mystic. The rooms of the sweltering hot second floor, however, were lined with altars and amulets- subtle to the untrained eye, as are many such items in the deep natural religions. Candles and bound sticks, bottles of rum, money that had been offered, plants dried and tied, and a few symbolic paintings were the only decor in an otherwise stark room.

The Babalawo led me to the smaller back room and motioned for me to remove my shoes and jewelry then sit on a simple chair on a woven mat. He sat at my feet. He mimed for me to write my name and birthday in a small book.

With Mechel translating, the Babalawo began the divination by drawing crosses on my open hands with chalk, chanting, and cleansing. He used a strand of coconut shell pieces and four small shells, each different, to divine for me. Sometimes he would cast them and record small marks in the book, sometimes, he'd ask me to hold them, shake them, or blow on them, then he'd cast them and write. Over and over he cast, counted, and made marks by my name. He wrote things, not in Spanish or English, but in a language I could not read. I asked no questions but tried to still my mind, steady my breath, open, engage, trust, accept.

When the casting was complete, he spoke at length. Of the many things he offered me, translated by Mechel, the ones that stood out were these:

1) My Grandma is always with me, protecting me. I need to put a glass of water for her someplace up high in my home.

2) I have negative energy attached to me. It attaches easily to me because I am *different*. I am at risk of bringing it home or into other people's homes. I need to cleanse my body and my mind, I need to banish these negative energies that seek me and attach to me. I need to take my shoes off indoors, shower before I eat, be careful about what I take inside my body, inside my space. An illness is imminent.

3) People watch me, are watching me. I stand out energetically and people look at me more than others. This is

dangerous. I need to learn to protect myself from this.

4) Someone at work is speaking badly of me.

5) There is a disconnection between what I do for a career and what is in my soul. I would do best to align them.

6) Twice he repeated that I need to be careful who I let close, keep close to me.

7) I need better boundaries. The negative energies that attach to me cause me anxiety and thoughts of suicide. I must not kill myself no matter what happens to me in this life.

8) I need to keep my sisters very close to me and honor those relationships.

9) I am being told to slow down, listen more. Listen to my life. There is a disharmony, my heart is not in my work. My life needs to come into alignment, I need to allow my heart to direct me.

10) Santa Barbara is my protector.

11) There is a baby waiting to be born to me. She will come soon.

After the reading he carefully wrapped two chunks of white chalk in a paper bundle with some of his hair. He told me to mix the chalk with water and clean my house top to bottom when I got home to rid it of the bad spirits who are harbored there. He said to physically throw the water out of my house when I was finished.

Then he asked if I wanted him to clear the negative energy that was attached to me at the moment, and of course I accepted.

He led me back into the main room of the second floor. An altar of black painted sticks bound in bundles stood propped in one corner. As Bex and Mechel looked on, he guided me through a ritual there.

He lit a small white candle, and using it to drop wax on the floor, he secured it to stand in front of the sticks. Chanting quickly and quietly, he took brown paper and wadded it into a ball, placing it on a short metal stand that held dried herbs and other small offerings I could not immediately identify. His chanting growing louder, he then took pulls of rum and spat them in a spray over the stand, the sticks, and the candle. He rose, and we both stood in front of the altar. Asking me to spin in a circle, he took the wadded paper and rubbed it lightly over my body as I slowly spun. He put the paper back in the stand, then, motioning for me to cover my breasts with my hands, he spat sprays of rum at me, his eyes rolling back as he yelled forcefully and incomprehensibly. Finally, he handed me a maraca full of seeds and asked for me to shake it at the altar while stating my intention. Bewildered and not quite in my body, what came through me is "I want love, I want to *be* love."

He chanted loudly again, *the candle went out*, and wide-eyed, I returned from wherever I had been to the strangely domestic setting of the simple room, where we had been standing in the whole time.

The rest went by in a blur.

I thanked the Babalawo, left my offering for the Loa on the altar, and put my sandals back on.

"He says take the wadded paper and throw it in the trash someplace outside," Mechel instructed. "It has in it all the negative spirits that were on you and in you."

I thanked the Babalawo again, carefully took the paper in my left hand, and with that, the three of us made our way back down the steps into the hot street.

"There." Mechel pointed. A dumpster sat in the middle of the throughway about a block down.

"Don't worry about putting it in there, the trash in Havana is full of this kind of thing." He laughed again, his infectious smile lightening the mood.

With an exhale, releasing all the personal negativity and dogma that I knew how, I ceremoniously flung the paper into the fly-swarmed dumpster. The contrast of potent ritual and broad daylight city life was oddly fitting.

Bex and I said our farewells to Mechel, customarily kissing cheeks and thanking him. Then we set out walking, slightly bemused, toward the apartment and whatever other adventures Havana had in store for us.

"*Pare, pare!*" someone in the back of the bus yelled.

The driver screeched to a halt in the middle of the narrow broken asphalt road. The hatch had popped open again, and all the luggages had spilled into the ditch as we rounded a curve. It was the third time. The man ran back to recover them while the rest of us waited, packed into the coach like sardines. Some people had resorted to sitting in the floor rather than stand with nothing to hold onto. I had a one-cheek claim on a bench seat for the duration of the six hour journey.

"Do you think we'll have our backpacks when we get to Trinidad?" Bex asked, laughing.

"No telling," I tittered, shaking my head. I had remembered to keep my passport and phone charger in the small pack on my lap. I flashed back to Lima, when I was so anxious that my bag wouldn't make it onto the same bus as me. Here, in the smoldering heat of central Cuba in July, there was no energy to spare for worrying. We'd find out in a few hours whether we'd get to change our sweaty clothes or have to make do. The potential for massive inconvenience had to be viewed as part of the adventure in this place. I smiled.

Looking out the front windows of the bus, I could see the tiniest glimpse of the ocean. We were almost to the southern coast. Palm trees fluttered in the hot wind like hands limply wafting us onward as the driver climbed back aboard.

The drums are a part of life here, and not a second passes that they don't envelop you from layers of distance and direction. They lend a seductive exuberance to this place, and your heartbeat changes to match whichever ones are pulsing closest to you. There is no escaping their spell.

Music puts us in the present more than any other form of human expression: it only exists in moments. And the moments here are very rich and alive.

I locked eyes with the gardener as we passed. Beautiful black eyes full of secrets, glinting in the already sweltering morning sun.

I wondered whether many people who stayed here actually *saw* him. After just a day here at this gaudy and grotesque monument to disposable income, my stomach was twisted by the unshakable impression that local people were invisible here, relegated to behind-the-scenes service jobs. No Cubans vacationed here. This was a place where people could come to Cuba without *really* coming to Cuba. Ugh. I should have done more research, but here we were, prepaid and checked in, no real town, just a gargantuan resort. We were marooned with the brittle upper crust of society.

The isolation of this place was geographic, but more than that: it was emotional, racial, monetary. It cut me off from the very world I'd traveled to Cuba to experience: the culture, the people, the varieties of daily life in their glorious honesty. I wanted all the messy complication or refreshing simplicity. I abhorred the clientele here at this monstrous complex in an island paradise, and I suffered internal wounds for my complicity in this brand of modern colonialism. I wouldn't have come here had I known it would feel this way, had I realized that the only places to stay in Cayo Santa Maria were all-inclusive resorts. These, for me, were the absolute bane of humanity and a paragon of classism, the type of place one goes to be insulated the realness that I was starved for. I wondered if someone with those soulful eyes could actually *see* me, separate me from the stuffy, entitled pricks that come to a place like this. But ironically, here I was, the same as them: apparently of the means to afford this hotel, but maybe privileged even more so to hate it.

Vacation is a thing that I no longer comprehend, as I have no need now to escape from anything (or maybe I'm on a constant escape from everything). But anyway, new places and experiences feel like they are *everywhere*, available for enjoyment. Most people don't think this way- I didn't always.

Somehow, when I went crazy I also started the process of freeing myself permanently. For those who work their lives away and get a set number of days each year for their perceived freedom, I can't imagine wanting to spend them here... but what the people staying here seem to want is *not to have to think*. They don't want to think of schedules or transportation or food or drinks or transactions or language barriers, they don't want to venture out into unknowns or meet Otherness. They want to be held in the soft lap of luxury to relax and indulge.

But looking around, no one here seems happy nonetheless. In the main building, people sit dressed in uncomfortable clothing, listening to elevator music, staring at the same people they already know. There are more frowns or neutral faces than smiles. A girl in her late teens in a white sundress and diamond earrings builds a tower of playing cards over and over, each time with the same perfect posture and look of vacant determination. She is a bird in a cage. Everyone is bored, but they've convinced themselves that this is vacation, and they've idealized that so much that they reject the emptiness of the fantasy. They all have the same cocktails with umbrellas and clap politely on cue.

The numbness of it all baffles me. They seem to attempt to find joy in escaping their grueling jobs and mundane routines, but in truth, they love the misery of the whole thing. They aren't able to access the freedom and happiness possible within themselves that they could have anywhere anytime, so neither are they capable of it on their vacations. They love to complain about the food, the service, the heat. It separates them and makes them superior to their surroundings. They take comfort in the familiarity of their collective negativity. The behavior I witness: the entitlement and treatment of the staff, is appalling.

As challenging as this place is for me, I am determined to dredge the depths of my reflections for meaningful truths. I am determined to leverage my personal discomfort into an opportunity to learn about myself and human nature. I am hellbent on making the most of my experience here. Some places are gems, some places are contrast. I am eternally grateful for both, or I'd never know the difference.

I think of the gardener again. I wonder what his life is like, where he lives, what he does for fun, and what he really thinks of tourists. It's a terrible mirror and an uncomfortable thing, privilege. It's a spectrum of many intersecting scales, on which we all fall somewhere. And while I don't wish away any luck I've had in this life, I don't find good company among people that *want* to feel separate. I've spent a lifetime fumbling to achieve inclusion.

"Let's just get to the beach," Bex sighed, her tone reflecting my thoughts as we skirted the man-made waterfalls.

Our disenchantment was palliated once we were at the shoreline. Powder-white sands met the crystalline vastness in sparkling resplendence. Looking out at the magnificence of nature here, I was at peace again. The real richness is always free. We staked our claim in a palapa, its dried fronds rattling tranquilly overhead. Bex settled in to sunbathe and read, and I headed out to grab a kayak.

The small vessel carried me silently, gliding over the aquamarine depths further and further from the shore. The communion with something Real was everything I had been missing there, and I was relaxed, relieved, and grateful.

I was starting to notice a pattern in myself, having collected enough data points to see it clearly. My healing was linked to feeling *connected*, and I was fumbling toward finding ways to both access it and allow it. I was reaching out into the world for kindling, and I was reaching inward trying to create a spark. Maybe one day the fire would stay lit.

Home Again

It never really heals,
I just get used to looking at it
Used to the discomfort of it.

Time passes between the lines
Measured oddly in days that stick together in the heat,
Pull like skinned knees stuck to the inside of long pants
When I'm trying to sit still,
When it's too hot to wear them but I'm told I have to
Be appropriate.

My body becomes the too tight pants
too small, too hot
I can't stay in there
Inside my own life where I can't breathe
Where the wound can't get air

Time passes and they say how happy I look now
Compared
They see that because they need it to be true,
Need to believe that time sutures these things
But all that's really changed for now
is how tightly I wrap it,
Bury it with a smile
So that I can give them the gift
of the more comfortable delusion.

What will I write
When the Healing has run its course
When the medicine has worked
And I'm whole and restored and alive?
What will I write to commemorate the time
When the Becoming is less severe,
More nuanced,
And gratitude has shifted its focus from
The relief of Surviving
To the creation of
little routines to
entertain or challenge myself again?

I don't take the time to *keep* as much
when I'm vibrantly present
And not drowning inwardly
in some passion or despair.

What a shame that it is
That I've immortalized
only the hardest versions of myself
And have yet to capture the small, incremental joys,
Far more abundant:
the unremarkable real victories
of breath after breath.

Pain is the greatest leverage for
creation and expression,
The walls between our awareness and
raw animal selves torn down
Our introspection pierces the fog,
illuminates the way out,
finds the way to return to equilibrium.

But what of these walls when all is well?
If happiness and love is our most natural state
Do we simply live them without reflection,
Fully enthralled in our existence?

This is not the ceiling, I think.

I want to access the flow of life
With the same magnitude of
Unrestricted framing,
The same urgency to achieve meaningful perspective
That I had when I was shattered;

I believe there are levels of
consciousness and love higher than this
That can be obtained
By trading my coping mechanisms
For thriving mechanisms.

I care about someone who deserves more than he is getting in his current relationship. His story hits so close to home for me, reminds me of the time when I deserved better from another, when I had given far too much time and energy and self. I had depleted myself into oblivion and reinforced an enormous delusion to find him meritorious. I remember with cutting clarity how it felt when the fallacy was laid bare. I was spent and angry at myself. I was destroyed by my unmet expectations of a partner who didn't sacrifice equally but expected the world of me.

I listen to this parallel storyline unfolding in its early stages and my heart breaks in embodied empathy, like his pain is my own. I know that same frustration, the burden of needing to make a difficult change, and the sting of disappointment in our closest people. I know the disorienting willingness to keep giving even when the pattern of imbalance is revealed.

His story also reminds me of all the times others deserved better from me: the times I couldn't see the impacts of my choices on people who loved me, couldn't understand the selfishness of my behaviors. It exposes the burden of a deep guilt that I still carry as though that humility will cleanse the record of my crimes. It dredges up the shames I still feel in my failures as a partner and as a friend, and the weight that the lack of forgiveness for myself has borne on my psyche.

Reciprocity is a complex concept because relationships are not merely transactional, they are a hall of mirrors for our conditioning and desires. When it comes to sacrificing for another, our ego would love to view our motivation as pure, our giving as selfless. But in the myriad reasons we continue to offer more up, particularly when we are receiving little in return, few of them are unconditional. We do have free will even if we don't always have conscious direction of our agency. Our subconscious needs can overprint our intentions and land us in dysfunctional situations. Our disproportionate contributions to a relationship can reflect our maladjusted attachment mechanisms or a distorted self-image.

In a healthy exchange, we give because we believe in a person or an idea so much that we freely offer our attention, time, love, or support. We give because we value the joy it brings those we care about. But even as pure-hearted as this can be, we often do it because we ultimately (even if subconsciously) think it will serve us to do so. We make sacrifices because we believe in love, believe that perhaps someone will sacrifice for us the same way, and in that shared giving we can elevate our lives. Therein lies the implied expectation.

Giving can also become twisted with obligation, a need to feel seen or acknowledged, or a need to play out power dynamics with others. Maybe we like to feel like caretakers, or we view others as helpless, or we like to be seen as charitable, or we have some familial or dogmatic obligation. We can give in attempts to make others feel bound, guilty, or loving. We can give too much because we enjoy our own martyrdom. Thinking back to why I increased my investment in a relationship when it was in its death throes, I realize that you can give in manipulated desperation as well.

Reciprocity at its highest is when givers give to other givers, and a brilliant, abundant balance arises. Even though we are giving because it brings *us* joy to see others happy, others are giving to us in kind, and an emotional elevation occurs. Relationships thrive when this equilibrium is easy to obtain, when people care and contribute without expectation or attachment.

The imbalance of reciprocation is destructive. When the scale of, or reasons for, people's giving is out of alignment, resentment is bound to follow. There is a sort of a karmic eventuality with those whom you exchange energy regularly. Over time, balance will be found at the lowest consistent frequency of the interactions between people. The one who puts in more effort will eventually give less or opt out. Most relationships suffer greatly in this kind of rebalancing, as the withdrawal of the major source of energy leaves too great a gaping deficit to overcome.

All of this reflection today made me think:

Am I meeting the energy of those people in my life that I care about in ways that feel meaningful to them?

Am I giving my time, attention, and effort to others in the same magnitude that they offer me? Why or why not?

Do I have any attachment or expectations on how I am received by those I care about?

Can I increase the frequency of my favorite relationships by investing more and offering more of myself?

Am I able to let go of people that seem to drain me?

Am I honoring myself by sacrificing only in the ways that I can emotionally afford?

Who is showing up for me, and who am I showing up for?

How easy it is for you
To believe you want more,
Want something of me.
You know me well because I offer myself up
When I don't know you at all.
Do you ever think of how little you extend?

"Did you write about *me*?" he asked quietly.

"Yes. But don't worry, I have fallen in love with increasing tragedy since you," I bantered in feigned despair.

"So I'm the one that got away?" he plied with an impish half-smile. That wasn't what I had meant. His face was older now, but his eyes still sparkled like a shallow lake in a late summer sunset. Lifetimes ago, I stared into those eyes in search of absolutes and tomorrows. How strange it was to be unencumbered by their spell.

"No, my old friend," I countered, smiling. I felt the odd fullness of the circle closing, the boundaries of my word choice, "*I am.*"

We laughed. It was bittersweet to see each other, to share our familiar badinage after so long. We had both changed so much since we dated in our early twenties. I was surprised when his request to see me came out of thin air. I wondered inwardly about his motive for reaching out. I couldn't gauge whether it was innocuous or if he was searching for something.

The thing about going back is that you never can, really.

Maybe you've come here to search my face for traces of the lover you knew, maybe you stand here in silent apology for the chapters of hurt we wrote on the skin of each other. Maybe you would like to show me that you are winning or see if I am, maybe you need to imagine how life in a parallel universe would have turned out. Maybe you just missed me.

But your reasons, while I hold them gently, don't ultimately matter because I mourned you completely. The girl you may think I am is a girl I wouldn't recognize now. It has taken me years of growth to sit with calm wisdom in my heart about past loves, what-ifs, and almosts. I don't see these stories as poetic dramas waiting to be finished, but as the soul lessons I needed in order to gain the strength to Become.

I have no doubt that Love, that beautiful electric thread that weaves people into some simultaneous narrative, is the only thing that will ever matter enough to change me. I do it again and again, kill myself over and over only to believe I can never recover my heart, until somehow I do. It's absolute, a full metamorphosis. In the course of loving and dying and loving again, there is a suspended pause when one must let go entirely and have no control, no consciousness, no captain... and really, no ship left either. There is the Nothing that must occur, and it must happen completely for the reincarnation of the heart. It doesn't matter how you feel or what you think in that pause. In fact, if you can still feel or think you haven't quite reached the bottom yet. You can't plan for it, fake it, cause it, or control the timing of it. You will only ever know you were there after you've made it through, when you've grown past it and come back together in a new way, when the cellular pieces of self that have survived the annihilation come back together by some gentle gravity on the other side of the black hole. It has nothing to do with your will or wishes, and everything to do with the process of Becoming, the evolution of a self. I have died so many times since loving you, there is no way to truly explain.

You re-enter my life like a long-period comet coming around yet again, glowing and aflame but confined to your forgotten orbit. I receive you with compassion and reverence for the way you catalyzed a massive destruction and rebirth for me. We were children then, in more ways than not- still we shared something Real that changed both of us. But I've lived a thousand lives in a different Knowing since I was the girl who fell in love with you, or the girl who remained in love with a construct of you for so long after. There is only nostalgic continuity between her and me, she is but a memory for both of us.

I look at you looking at me, and I know that you haven't done the assignment, haven't ever been able to let go all the way. You haven't gone through the same hell, the pause, the renewal. You held yourself together and stayed still. I look at you and feel a million things: empathy, sadness, gratitude, love. But not the kind of gripping, life-Love that can really touch me or change me again, not the kind of Love that could rearrange my DNA or upheave my soul.

It's not a rejection in the least, nor a judgment upon you. In fact, I hold you safely in a special place in my heart reserved for those who have altered me profoundly. I couldn't be who I am without You.

I did write about you. Of course I wrote about you. But I can't save you, and in a way, whether you see it or not, I realize that's why you've come, why your soul has sought mine again. You didn't level up, you would like to try once more, you need me to destroy you in love. You believe that I am capable of this. But I'm not your impetus, you will have to find her. The most compassionate thing I can do now is to release you completely to continue your path. I honor you, I love you, and I wish you every luck.

If you ever come looking for me
Don't look behind you
Where you think you left me.
From the moment we parted ways
I found I had a lot less to carry
And could run free and blithe,
As light as dandelion seeds on a breeze
As playful as shadows cast through maple leaves
Feathers dancing
Hair flowing
Fingers running through the foxtails
In fields of sunbeams.

Now that I think of it
Just don't come looking for me.

In our full moon yoga practice last night, lying in savasana, a question was posed: *what is pulling you away from being absolutely present in this moment? What is taking you out of this room, out of this experience, out of simply being?*

I want to look at this honestly and without self-judgment so that I can try to find ways to meet the needs that are distracting me from living with greater presence.

Here are my major sources of mental escape or disquiet:

1. Things I believe I need to do, plans I've made, packing lists, places I want to go, things I anticipate experiencing.

2. An attachment to validation from social media and messages, thinking of people I care about who are elsewhere.

3. Rumination on past interactions, past relationships. Pain from past rejections, confusion from past emotional abuse, attempts to understand and reconcile past occurrences.

4. An attachment to image. Clothing I'd like to buy, gear I'd like to own, imagined scenarios where I'd wear or use these new things.

Why are these my distractions from an active participation in my current life? How can I give myself what I need to resolve these sources of discontent?

1. It's funny to admit that I'm not able to be present because I'm daydreaming of other experiences I will have *where I probably won't be fully present either!* I remind myself that there is time to be here now. There is time to be there then. Planning and packing and arranging the calendar to maximize the breadth of experiences is important- but not as important as fully engaging the things I plan. Just calling awareness to this helps me understand how I can make adjustments to improve it. I am going to add some short meditations to my days to bring me back into the moment.

2. Lying here in mediation, what's pulling me out of my experience? What's tugging me into conscious thought? Ugh. *I'm thinking of checking my phone.* I'm wondering what's going on with my friend who just had a baby, and whether I'll get a message from the man in La Paz. I'm wondering if the friend I traveled with in Cuba is over her heartache, and whether my kindreds in Costa Rica are thriving. My friends and family are scattered all over the world, and I love that I have a device that connects me so easily to them. But on deeper reflection, I'm lonely, and I've become attached to the idea that a pile of messages awaiting me can alleviate that. My ego is involved. I crave validation, I want to be seen, I want to be adored. I want reciprocity in my friendships, I want to invest and have people invest in me. Only... I feel disconnected in my immediate environment, and I am relying on people far away to create and share intimacy. What I realize is this: while the love I share with my remote friends is real, my reliance on the electronic proxy to feel close with other people is less than ideal. The answer to my loneliness is not inside my phone. Those messages that await me, or don't, won't change what's really lacking. This one is not a simple fix. I need to cultivate deeper relationships in person, and I need to start by addressing what's blocking that. Is my inability to get close to others at home a means of protection? Am I lacking people around me that share the same values or perspectives? Maybe a break from my phone is in order. Will I think about it more or less when I can't check it?

3. Ohhh the past. I'm getting *relatively* better about not allowing those memories, nostalgias, and things I should have said to steal my time, but it still makes the list. Part of the reason I think so much about the past is healthy: it's reframing and learning and healing. But partly it's obsessive and self-destructive, and that is harder to admit. I dwell on the mistakes I have made, I worry about how I have hurt people. Reflecting on this, I see that the way out of this issue is forgiveness and integrity. Moving forward, I resolve to actively continue the work it takes to forgive myself and others, and to better center my actions with my high self. After some

time passes living in alignment, free of new traumas, surely there will be less to inspire these painful ruminations. As people resurface into my life, which seems to be an odd pattern emerging, I will meet them with compassion and boundaries.

4. Attachment to image. This one is easily the most embarrassing to admit. I'm sometimes pulled from my present moments because I'm daydreaming of things I want to buy, things I want to wear, places I'd wear them, gear I want to own, how much fun I could have using it. I am wishing I was elsewhere, and a little movie reel of me looking radiant and being happy plays in my mind. This happens most frequently when I am at work, or if I'm feeling especially bored or discontent. I am not shopping for a dress, I am imagining myself at a really exhilarating party. I am not looking for a new kayak, I am longing for the way it would feel to be on the water with friends. It's the emotional component of advertising that has me on the line.

But it's not just escapism, some vanity is mixed in too. It feels gross that I am stuck in this socially-conditioned low-frequency game. My materialism is, in part, a coping mechanism I have from growing up poor and getting picked on by classmates for wearing yard-sale clothes. The lack of feeling accepted in adolescence sparked a subconscious drive toward having all the finer things once I was able. Those things provided an insulating buffer between the cruelty of the world and that ashamed child still within. My "enoughness" was commoditized early, and I want to reclaim it. I was unwittingly subscribed to the system of an endless chase, where the beauty and belonging we are sold never arrive with the packages.

Buying things has become a bit of a drug for me, as confirmed by the fact that I'm lying in savasana thinking about getting a bikini to match the beach pants that I want to bring to Panama. The movie plays, and I see myself there on a beautiful white-sand beach (escape?). Sunkissed and glowing, I look serene (is external beauty being confused with health, inner peace, belonging, and contentment?)

The misconception in all these idealized little daydreams is that the times I have been the most happy in my life, I couldn't have cared how I looked at all, and I truly barely noticed where I was. I have been the most full of joy at times that I shared with other people I love. A common theme is revealed: the root of this issue is from lack of connection as well.

The things keeping me from engaging my life in presence and flow mostly come down to 1) a need to connect deeply, 2) a need to forgive (mostly related to failed connections), and 3) a need to live with more integrity- in my career, in my relationships, and in my choices.

The ways that these things keep boiling over in my life are the source of, and therefore the solution to, my problems. Awareness is the first step toward the remedy.

And this is why I can't drink
As it raises the dead
The feelings about you
and about the world
and about loss,
About here versus there,
Then versus now...
The thousand yard stare that I can't help
Because a thousand yards is as far as I can see.
You're 1500 miles and two lifetimes away.

Through the open window, summer.
Through the window,
the bass riff of distant music
that I can't hear the rest of.
Through the window, all the lives I'm not living.

This is why I can't drink
This is why I can't sleep
I'm hung in the balance between worlds
Understanding both and neither,
Belonging to both and neither.

I forget sometimes how much work I've done on myself out of survival, and how much I must have changed. My attempts to bring him into my current world are anxious, vulnerable, frustrating for us both. The gap between us is a cavernous throat yawning open, lips impossible to bridge with a kiss. I suffer such a great loneliness that even in offering myself completely I can't close the distance. I'm misunderstood. I can't stay home for long before this wounds me, so I know we still need more space for perspective.

I read to him the insights about what pulls me out of the present, and I was surprised when he looked injured. He is a fixer, I must remember. And I am broken, but we are both without hands, without tools. I try to explain that I am not unhappy, and anyway he is not responsible for my happiness. I'm one of those people who dig deep to reflect this way so that I can optimize my life experience. Nothing is wrong, I'm just existential by nature. I'm not perpetually displeased with the state of things, I just try hard to see it clearly.

He views me as a person intensely upset with life, having so much she wishes to change. For me, confronting the things I can improve about myself is positive, healing, important. I have been through hell and have learned that only by facing your monsters can you tame them; only by starting where you are can you create substantive change. You have to be willing to admit flaws, sit with discomfort, leverage mistakes. It's constructive in my opinion- the path to growth. I thought he'd be proud of my introspection, but our frequencies are discordant.

I am attempting to train my mind, tame my emotions, and level *up*. I am rewriting my warped programming. I am getting unstuck, I am learning to swim! But in all my thrashing to keep my head above the water, by now everyone else just wants out of the pool.

I've just experienced my first mixed bipolar episode since I've had the words to apply to such things. For the last week I've had trouble sleeping, crazy dreams, rapid cycling of moods, and overlapping disparate moods. My hypomania has manifested as faster talking, boiling excitement, planning, fast processing of emotions, and an overwhelming desire to communicate my feelings. I can't concentrate on work. And I've been writing *a lot*.

The depressive symptoms tangle up with the stimulated ones. I'm having palpitations that feel like panic attacks. I'm sullen and lack energy- unless I have too much of it. I can't eat, or I binge. Nothing interests me. I daydream of traveling, running away, or committing suicide. The only one of these I plan for is traveling, but it doesn't bring me relief or joy. It feels stilted because I'm desperately lonely, and my last two trips didn't bandage that like the ones before. Planning feels dangerous because I can't connect to my emotions properly, I can't currently process whether my behaviors are self-destructive. I can't realistically gauge what I will feel like doing in these places because I can't imagine this overbearing emptiness subsiding. So... horseback riding or jumping off a bridge? Tours where I'll meet people, or a quiet room so I can write all night?

I feel trapped because there is so much I want to do, and I can't make myself do anything. I feel unseen, empty, and ineffective. I have strange headaches on the back of my head that I've never noticed having before. I take naps to try to reset. I have the feeling that I'm crawling out of my skin. There is a war going on inside of me and I'm not on either side: I'm the land on which it's being fought, I am the opposing philosophies, I am the innocent civilians. I am struggling against drowning. I feel the overwhelming urge to communicate coupled with the severe sadness that there is no one to talk to, no one that can hear me. I am screaming into a vacuum, I am whispering into an empty house. I could post all my feelings online, or publish them in a book, or call and tell all my friends, and I would remain impotent, invisible, lost in translation. Exposure only generates platitudes and pity, it never creates the space I'm silently begging someone to hold for me.

The experience itself combined with my depressive thoughts about the experience is isolating.

I also just put it together that I write exponentially more when I'm hypomanic. So much makes sense in this context now. I edited one book and cranked out a whole new one in the six months I was high from antidepressants. My writing has slowed since being off the drugs, apparently except when I have an episode. When I'm in this state, I feel like I can finally connect all the dots in the cloud I otherwise live in- the ideas pour through me, and I can't write them down fast enough. The drive to make sense of the enormous amount of sensory input and emotion breaks the barrier between my subconscious and my fingertips, and I compulsively commit my elucidations to paper before they are lost again to the ethers.

I looked this up today and there's a word for it: hypergraphia. And it's absolutely linked to bipolar disorder.

I'm a darker black
Than the rest of the night
And it doesn't try me on my way home
Street lights straining
To cast bigger shadows off of me
But I throw depth at the ground without them,
Need nothing to find my way to somewhere else

I overstayed
And I pour my black into the guiltness,
My guilt into the nothing,
Because it doesn't matter
how empty it seems walking away-
It's emptier within.

Why do you stay? Your question cuts to the heart of the thing, and leaves me at a loss. You see my restlessness as symptomatic, an indication of disease in my relationship. The phone signal crackles in the pause. The one thing I won't hand over to you is an analysis of my marriage for your scrutiny, so I deflect your probing. But later, in the safety of my aloneness, the question permeates my thoughts.

The truth is there is nothing *wrong*, so of course I stay. I love him. I love our life, our house, our friends, our routines. I love what we have built together, and I don't imagine ever sharing this much with someone else. We are best friends, epic partners. We have been through hell and back more times than I can count on one hand, and we will get through this one too. He and I will outlive the part of me that is indulgently infatuated with your darkness and dancing with my own.

The truth is, below the surface, an insidious little wound hides away, never really healing but never becoming infected. I don't *feel* enough, there is a numb spot where connection could be. I don't feel passion, and I crave the madness of it. I want someone who sneaks photos of me and calls me pet names. I want someone who asks what I'm writing about and why. I want fights and breaking dishes. I need electricity and fucking and the kind of explosive emotion that sets off the car alarms outside. It loops within me with no outlet, stuck, and occasionally I do wonder what life is like someplace else. But what life do I imagine that I'd actually leave to find? And do I think it's *Real?* I stay because it's not his fault that I am wracked with wanderlust, confusing intensity with vitality... yes, you can feel more alive by sticking a fork in an outlet, but surely it's not the only way.

I stay because the issue is within me, not between us. I am working through trauma. I'm too damaged to feel at home anywhere. I stay because I don't believe the connection that I idealize in the most secret, locked-up reaches of my heart really exists. I have confused love with some romanticized escape where someone else saves me from myself. Emotive, mad love has a short half-life and doesn't exist for long in the real world. I stay because some short-fused passion isn't more important than everything else, and literally Everything Else is what I'd give up if I left. There are no fights, there is no drama. There is just this patient, selfless, grown-up domestic love that I'm too broken to accept properly at the moment. It's protective without being parental. It's kind without being smothering. It's the most pure, perfect love... it just lacks a parallel for the ways I am insane. It's lacks the abuse that makes me feel appropriately punished. It lacks the desperation and tragedy that are apparently the sick ways I am conditioned to feel anything. *I* am the thing in disrepair, not my relationship. So I stay because now I think maybe I can heal. I stay because I think I can fix myself, and maybe by some grace or mystery he will still be here. I stay because he gives me hope for a future.

It's dangerous to deepen our conversation to the level where this kind of admission resides, so I won't. Because you are broken or awake or mad as well, because you are asking to see the cracks in the foundation of the house I built, and surely you have your own agenda. If we talked about such things, you'd twist my words and find purchase on an opportunity. My loneliness is so slippery, and I am leery of your extended hand. If I admit to you that I don't feel enough where I am, you will use it as a crowbar into the locked cellar of my heart. Soon we'd imagine that we could rescue each other, even though it's impossible. There is no outrunning a delusion that would fall apart, crumbling from small realities at first, then eventually cleaving from big, stark, unapologetic ones. The excitement of irrational possibilities waits to sling us in its gravity around the hold of the black hole if we let it... but I have seen this before. We'd only burn each other alive to come out the other side expended, scarlet-lettered, and reeling.

It's day 13,278 of my life and I'm wasting much of my time now in the closed room of my own subjectivity, with the fear that someday I will forget how to find the door out again. When I am in this state, I have a sort of emotional synesthesia that takes hold, some kind of relentless desire to make connections between the inner and outer worlds. I have to write, eat, love, and touch *everything*, and I have to do this because I *feel* with either impossible intensity or numbness. I am exploding past sensory confinement. I have to experience everything anew, as though I'm a child and a channel between worlds. All the portals to my senses are shattered open and unfettered. I'm enthralled yet inept in this overflow, just trying to keep up with all the input coming through me unfiltered, raw, and real. There is such urgency to live expansively in these highs, because they are not promised or predictable. Only when I feel too much can I feel at all right now. How can I ever explain?

There is a great cost to the highs, but it's not as though I choose to perform these transactions. I'm being robbed and going for a ride in a trunk, but the sickness tells me that at least I'm not bored. The headaches and partial numbness, the impulsivity and compulsion, the blurred vision like looking through a kaleidoscope. I'm shaky and can't bear to be around people. Or I'm low and secretly feeding the dark desire to self destruct in private as soon as I can. Or I'm excited and loud and creating events and planning travel and drinking too much, talking too much, racked with insomnia, and exhausting myself in fits of enthusiasm.

What a bizarre disorder my brain has chosen for me. Did I do this to myself? By taking drugs or too many antibiotics or having poor stress coping skills? Is it from poor nutrition or a virus or shitty genetics or trauma? Am I reinforcing the disease by not being able to fight off the episodes? In some respects it's an asset, and in so many others it's killing me, fast or slow, or both simultaneously. I am only now barely able to comprehend how real this is. I can't pretend that I am normal any longer. I am afraid.

Every day I wake up and I fight for my life.

Every moment of joy that I collect awkwardly in tired arms
is in the hope that it can kindle a bigger fire

A fire that will stave off the wolves that wait
so hungrily in the shadows.

Growth sometimes means becoming less complex, simplifying, reducing, centering.

The nature of growth is cyclical with leveling up... increasingly complex as one nears the test, the point of inflection... then as one passes the threshold into the next level, energies shift once more to a simple state, breaking through to new truths that will again be built upon with intricacy as the cycles continue. Our growth is electron shells, every next level requiring even more energy to transform Self into New Self.

I said that I want an optimal self and an optimal relationship. I was challenged with the idea that this word implies completeness, a sort of upper limit, an end. What I really mean is that I want a *thriving* self and relationship, so I brainstormed on exactly what that means to me...

Thriving self:
- sees the world with clarity
- is introspective
- feels gratitude, hope, and optimism more often than not
- actively tries to increase the love in the world by increasing the love within the self
- notices mood changes or illnesses, and takes care of the Animal until mood stabilizes or illness abates
- works to restore mood balance, mental, and physical health
- finds ways to love others the way that they best experience love
- nourishes flow experiences
- values relationships and experiences over money, image, or luxury
- contributes positively to society
- has healthy coping mechanisms
- is present-dwelling (more than past or future)
- creates healthy interpersonal boundaries
- is deliberate with time and energy
- can say no with grace
- manifests a loving support network/ chosen family
- recognizes impact on others and has accountability
- acts with integrity (at physical and mental peace with thoughts, actions, and interactions because they align with the high self)
- forgives self and forgives/ releases others
- levels up in consciousness, gives energy outwardly
- growth mindset but stable sense of self
- has presence

Thriving relationship:
- communication/ active listening
- attention and consideration
- sense of growing together and as individuals
- interest in the ways the other person is growing- actively getting to know each other as we change
- stable (neither conflict-rich nor conflict-avoidant)
- respect for each other's autonomy, personal choices, personal freedoms
- shared sense of humor
- shared sense of adventure
- mutual desire for health
- specialness (rituals, dates, messages, priority)
- sense of closed inclusion (things only we share together)
- loyalty (sense that no one can come between us)
- healthy/ playful sensuality, flirting
- physical touch other than/ in addition to sex
- kindness
- spontaneity/ surprises
- shared interests/ hobbies
- both partners give more than take
- mutual forgiveness, present and forward-thinking
- honest, vulnerable (and honest with selves)
- happy for each other
- taking care of each other, counterbalancing, knowing we can rely on each other

Follow your heart, they say.

Well, some hearts are easier to follow than others. My heart is dark and untamed, running wild-eyed like a feral wolf between the depths of belonging and the ecstasy of freedom, but never settling for long enough to make a home in either.

I have followed her full speed straight into white hot fires, unsure whether she started them or just offered me up as fuel. I've loyally revived her from ash, retrieved her from thieves, and resuscitated her from smothering. I've pulled arrow after arrow out of her until I learned that it hurt her less to leave them in.

I've also betrayed her. I've turned away from her when she lay bleeding. When she didn't agree with the plan I had made for us, I stifled her and tricked her. I forcefully bade her to submit. And for awhile she complied. Then she just stopped talking to me.

I owe her unwavering fielty for her unflagging hope, her warrior boldness, her steady persistence, and her childlike curiosity. I owe her now for the times I didn't follow her, didn't trust her, couldn't hear the things she whispered so she set me up to feel them, to scream them in my own voice.

Now I am still for her, I am learning to listen for her signals and read her signs. Now, she leaves me tracks on purpose again, knowing that I'm never far behind her in the woods.

Where will I go when I run out of places
To run between
And it's time to come back
And sit?
When it's time to go within
To the real unknown,
The last place left to explore?

The world has relit the torches that were
Snuffed and lost to ruin from war,
Refilled the cup
Long spilled of the hope of quenching just thirsts.
The sun shines over an aftermath that asks to be seen fully,
That asks when I am coming Home to rend and rebuild.

It is time to honor every heart that made a signal fire
All the hearts that stood like quiet lighthouses
while my soul pitched with
fevered delight that I may die at sea.

I'll plant a garden to name them all,
A place to bury all my selves leading parallel universe lives
A cemetery of living headstones for the ways we lost,
The ones that had to go beneath
so that we had shoulders on which to stand.

It's time to watch the sun rise on a land
I've seen many times but never felt.
Still with knowing,
Still with holding space like cupped hands to
cradle the fallen bird.
No longer the broken home
But the earthen nest for my heart of new roots,
Hungry and reaching to drink the meaning
Of a different kind of growth.

Selve: the act of become a separate self.

Witnessing the liquid of essence, of soul, selving
Individuation from the plasma of the collective.
Sacred entropy
Stratified consciousness layered by density.
I am plural,
Parallel to myself,
Woven in strands overlapping.
The texture of paradox,
Contrast that never finds the median
But touches in the moment like layers of music mixed;
Love is it's own state of matter.

Hunger like hands held too tight
Now emptied;
Hunger now a knife blade and no longer a question,
A heart that reaches into space with no arms
Phantom love pain
While nothing pauses or reaches back.

Estes Park

An attachment to this consciousness keeps us afraid to die. What if we never experience another thunderstorm rattling the windows in their panes, what if we never freefall in love again, what if we never again see daffodils unfurling defiantly against the clutches of late winter, what if we never taste another homemade birthday cake made in our honor, what happens when we can never again experience, *never feel anything* anymore?

There is such stark emptiness in this thought, the recognition of the void into which we will all one day fall, the time that we will spend without bodies, without reference, without senses.

There is meaningful contrast derived from imagining this. While we are here, we are alive to *live*. We are alive to touch, to taste, to hurt, to love, to make mistakes, to ponder our miraculous existence and our beautiful death. We are alive to find stories in clouds and constellations, to inherit the myths of the archetypes. We are here to be overwhelmed by nature and art, to grapple with the perfection in its design and reflect upon it for parallels in our constitution. We are alive to feel. We are born into these limited carbon vessels to have the chance to explore ourselves, to express what resides in the soul, to manifest love in a tactile way, and nothing more. Nothing could *be* more.

Where do you put the "almosts"
in the map of your heart?
The ones that were almost your sisters,
Your second set of parents,
The ones you almost ate Thanksgiving dinner with, forever
And called on their birthdays

Where can you keep them
When they can't be a home for you?

I wish our sense of collective inclusion
Were stronger
So that we could keep our almost-families close
And we could
belong to each other
With or without absolutes.

Even since I was small, I have often felt an underlying thread of disconnection within my family. And this wounds me subconsciously, goes against my idealized notion that I *should get* to feel close, understood, and held by my people. I want slow time, inclusion, and messy togetherness. I want a depth to our perceptions of each other and a shared security in our interrelatedness navigating this life. But we have separate villages, separate houses, separate world views, separate accounts of the same stories... and when we get together we seem to collectively lack the skills needed to actively relate on a meaningful scale. There are decades of injuries ignored or brushed aside, self-absolution in place of accountability. There is an adherence to our preconceived constructs of each other instead of an openness to learn. We are good at having a boisterous good time, but when it comes to anything deeper, it's categorically absent. They are the people I love the most in the world, but the ways that they are still asleep are triggering. "If you think you're enlightened, go spend a week with your family," Ram Dass once said. My heart aches from the truth of that reflection as my coping devolves, and I fall from expansiveness once again into the chafing role of the unseen child.

It's not in my nature to float on the surface for long, so I typically feel like the black sheep: separate, unheard, misunderstood, overlooked, and sometimes even judged. In a group setting they talk over each other, escalating in volume but not in content. They listen to reply, not to understand. There is little in the way of vulnerability or authenticity. It's a contest for attention, and the ones that aren't enthusiastically competing are dissociating- myself included.

My disappointment arises from the contrast between what I witness and what I know is possible. I romanticize an idyllic family life just as much as my personal relationships, my friendships, and my career, I realize. I ache in the gap. I want more.

My family as individuals are each incredible people- it's no wonder that I want to share deliberate, expansive relationships with them. But the space between *what could be* and *what is* hangs me up. In the chaos of our interactions I become ungrounded, frazzled, and isolated. I long for sincerity, reciprocal attention, and heart-centered consciousness. I want to show them what I have learned about relation-ship and holding space, but these things vanish from my purview among them. I want to actively listen to them, but when it's everyone at once, my energy gets overwhelmed and my focus falls apart.

I sit with all of this to accept it, but first I have to understand *why* I feel it, so I can truly let it go.

Maybe I feel irritated with them collectively because none of them walk softly or wait for someone else to finish speaking. In a group, they don't listen to each other or ask the questions that are important. None of them, even in their well-intentioned opinions, are gentle. Everything feels anticipated, structured, forceful, masculine. I flounder in this more than ever, because I feel outside it now.

It has become apparent exactly where my sensitivity to sensory overload began, and where my wounds from feeling unseen first formed. And as healing from those things catalyzed a life path of deviation from my family's behaviors, I became even more separate from the connection with them that I longed for. I sought higher consciousness, ways to feel like I belong somewhere, ways to make others feel the belonging I was so missing. I am frustrated that I can't *give* this to my family- I can't seem to show them or teach them this part of me despite my attempts. There lives within me an emergent, encompassing love that has connected me to a chosen family all over the world and quite literally saved my life. I have learned conscientiousness and softness, but these things only make me ineffective among my ebullient family now. I have learned that my sensitive nature is not something to be cured, but it's still a disadvantage here among them. How can you teach someone the wonders of stillness when it is not in the scope of their value system? They see things how they are, I see them how I am, and we've never seen the same things more differently.

I have to try hard to prevent myself from detaching in the cacophony. The tendency to protectively shut out the brash or cursory rears its head in my overstimulation. I have a hard time channeling my energy into perfunctory exchanges- still, I try. I can't possibly keep up with them at their frequency, and the rejection injuries of my inner child resurface as the family patterns play out in continuity with my decades-old imprints of insignificance.

Perhaps I become anxious because the disjointed garishness of our togetherness can be a reminder of ways I feel stunted from lacking familial support for so many years of my life. *Who could I have been with attentive, evolved, and open-minded parents and a stable home life? Would I have been able to develop more relaxed attachments without the churning fixation on first feeling more secure? What could I have achieved with a solid emotional foundation on which to stand?*

My criticism of them is too sharp, and it jars me back to compassion. It's not their fault, it's just where they are in their journey- just as I am at a vastly imperfect place in my own. We are all doing our best to survive and connect, despite the deck being stacked against each of us in so many ways. I want to let this go, let them be precisely who they are, and honor them without condition. I want to forgive their inability to relate deeply even as it hurts me. This is a lineage wound with no one to blame, but generations yet unable to overcome it.

Maybe we are all trying as hard as we know how. Maybe my family sees me more than they can express. Maybe my relentless draw to depth makes them uncomfortable, and I should be the one trying harder to meet them where they are. Maybe if I can *embody love*, I can show them the better way eventually.

Later:
I talked to my sister on the phone. She said she was also desperately missing a deeper connection when we were all together!

How can we all want something and not be able or willing to create it!? How can we all suffer for the degree of superficiality in our interactions but still be too self-protective to manifest the closeness that comes with vulnerability? How can we break free of our patterns to do something new? Humans are so complicated.

I can see now that for all the times I have succumbed to despair, aching to end myself, that I wanted to kill the false self, not the whole self. For so long, I had no idea that there was any distinction. In the nick of time, one pause, one blink of healing preserved the course, and everything past that has been borrowed from a miracle. The threshold was reached where the medicine tipped the scale, though the sickness was far from over.

My fever broke long enough to internalize the teachings of existential observation and glimpse my own construct of myself as separate and changeable, which spared me from my own hands. She was the work of the blind artists, a puppet on strings of subconsciousness. Her reality was the manifestation of rejoinder and feedback, stratified from the universal truths. The self confined to a life of pinballing in perpetual reaction *did* have to die, but I did not have to die to end her.

I came through the trial to find that it has only begun. I am shedding selves like winter clothes as I step into the warm room of liminal consciousness. I am not my memories. I am not my injuries. I am not the crimes committed against me, nor the ones I have perpetrated. The mystery seems bottomless, the removal of each mask reveals one slightly more Real beneath it.

Despite our clumsy humanness, there is a goodness at our core. The selves each so dutifully serve us as fully as they can, innocently interpreting through the lens of their unpolished predecessors. We cannot know what we know before we know it. The objective of growth is not to spurn spent selves like ignoble things to be cast away, but to understand what gifts they have given us, what indelible purpose they have fulfilled. In forgiveness, there is freedom.

Home

I feel embarrassed and almost... fraudulent that I am not further along in my healing. I wanted to write, to catalog my journey out of this depression. But while the overall trajectory is positive, there seems to be no real end, only the oscillations of temporary relief. The lows are less frequent but still *brutal*. I am not losing hope for absolute recovery, but I recognize that I have to keep journaling only for myself and abandon the expectation of sharing my experience. I have to keep my thoughts potent by protecting them from exposure to judgment, or worse, indifference. I have to cocoon to destroy the urge to present a mirage self. I have to spend the time internally where the magic lives with the ugliness of healing. It can be no other way. Writing about healing is easier retrospectively, when you can wrap it all up neatly because you know how it ends. When you write from within the gut-twisting grit of it, you run the risk of cleaving the private self from the public one in the defense of the image- faking or embellishing your real progress even as you objectively fight against doing so.

This is not a story of bravery as my ego would love to tell it... it is a tale of *my own survival*, and survival has a small circumference. From within it, there is no access to the means to make it beautiful. I've killed things and pounded my chest and screamed into the void and cried until I threw up. I have lied and hidden my thoughts and used people to feel better. I have run away, I have hurt people. And all these things are part of my process, part of this wrenching unfolding. Every goddamned thing I did, I did it to survive despite myself. Mine is not a success story, it is not a prescription for how to live or love or how to travel the world, or any kind of advice for someone else to heal. If I was ever brave it's because I had to be. If I ever seemed fearless it's because the monsters I had to fight were in the shadows, eating me alive from the inside. God, how I want to edit out the worst sides of myself, but if I do I have failed the test, I won't have created something Real. And still, the grotesque manifestations of self are not done stepping into the light, and I am ashamed of some of them, but there has to be space for their belonging as well. Remembering how I felt and how I got through this has to be kept pure, an accounting. Truths are easy to tell when they are simple, integrity is easy to showcase when you are already whole… but that is not the story of my Becoming.

The inflammation of ego will not resolve by indulging it, it will resolve when I cede control. I remind myself that I'm writing because it's my outlet to connect with the complexities of my inner world, it's a private indulgence for my own self-understanding. Whether the deepest scars on my psyche could be revealed to others is something I will have to decide later. That way, I keep my practice free from the poison of my own vanity. I've been blocked from writing more lately because my mood has shifted back to overcast, and I desperately wish my path was a straight line with landmarks of success, each chapter a little brighter. But I have to keep going nonetheless. In my life I will surely face other dark nights of the soul, when I will need evidence that I can endure. These journals are an alchemical manuscript with the exact equations to turn shit into gold. I can feel the transformation even if I can't see the results yet. Can you really write this from inside the chemical reaction? The sorcerer's stone is encoded here, unrefined and ordinary. The works retain all the wrong steps, the failed experiments, the erroneous thinking. It's critical emotional science to leave it whole, because the arrival at a solution surely requires the process of setback and perseverance, fidelity to the idea of healing. It's the hidden x in the algebra of the heart, but though I know this, the trials are painful. My healing is a house of mirrors and I'm so tired of looking at my own face.

I understand now that as I recovered from a broken heart, from emotional abuse, and from an identity crisis, I unearthed more wounds I hadn't yet touched. These things have been locked quietly inside me for years, background programs running in my psyche: the mother wound of abandonment, the father wound of trying to please others to feel seen, a disconnect from the feminine, an enormous gap between my soul purpose and my career, the ways I have been romantically maladjusted, my self-destructive tendencies, the damaged people I attract because of my poor boundaries, and the collateral wreckage that comes with them. So many things rose to the surface to also be acknowledged and healed. And I have worked to repair as much as I could, but it is truly a process. These things may have long been injuries, but it was only as I began to want to live again after losing my will almost entirely, that I could see the systemic damage. I have grave wounds from childhood and malformed attachments, I have a lifetime of violences I have inflicted upon myself, and I have coping mechanisms I developed as protections against feeling the pain from all of it. It's a cycle of energetic sickness and stuckness.

The acute healing from codependent love took everything I thought I had. So to uncover so much more work yet to do has been disillusioning. I fought for my life to return to this relatively functional state, the one that I have long since operated from, only to find it unsuitable as a foundation for the life I really want.

I think I suffered a relative loss of purpose after the first major round of healing, which tripped a sort of secondary depression. Once my will to live was reignited, it became my life's work to rebuild myself... and through therapy, travel, classes, yoga, meditation, writing, and learning how to trust again, I succeeded. I survived. I reframed, unlearned, rewired, reinvented, and renewed. I came through this changed, strong, and even more authentic, intuitive, and grateful. I never thought I would make it, yet here I am- every morning I wake up safely and happily autonomous, open, and alive. But *now what?* I am restored to a passably operational state, but my tenacity has dissipated in the relative lack of contrast. The new work is learning how to thrive without trauma as an impetus. The vibrance of life I felt so intensely when I was combusting into survival has been muted into routines and micro-victories. One cannot stay in a state of exuberance so heightened, just as one cannot stay in the state of despair that proceeded it.

Now my challenge is more subtle but even more critical: in order to achieve the next level of consciousness, I must find joy and purpose in smaller and smaller things. I must learn to put trust in the things that I avoid out of pain: domesticity, home, gentle love. I must patiently build quality relationships by remaining open, while lovingly cutting away relationships with people who are not interested in growth, reciprocity, or soul evolution. I want to be surrounded by people who make me want to be a better person. I must increase the integrity with which I conduct my life. I must work to repair the root of the issues that caused my attachment to a toxic relationship in the first place. I must carefully, and one by one, examine all the parts of myself that need restoration and attentive care to become whole.

None of these things are easy. I think they will be even harder than saving my own life from the major depression, actually. Persistence, perseverance, resilience…. I ask for these things coupled with clarity. *May I be deserving of, and maintain a network of souls that keep me growing, keep me open, help me on my path of leveling up… and may I help them on their journey as well. May I accept exactly where I am in my healing and take up the space that I need. May I take care of my soul, my mind, and my Animal with heart-open kindness, forgiveness, and support.*

I'm afraid that I'm healing only part of myself. The part that wants to be Love is becoming dizzyingly close to Source at times, but it hasn't brought the dark part of me up to ascend with it. I feel cleaved apart again, and the selves are more disparate in nature than before. One is an intricate paper kite, one is an immovable iron anchor. There is tension between them.

I can never tell if when I notice these things I should attempt to repair the dis-ease, or if I am just to observe, that nothing is wrong or changeable anyway. I never know why I see so much, feel such extremes. Am I broken or am I evolved? My own introspection is crippling and self-referential at times.

I saw the dark heart again today, the one that stays empty. She has not been fed with the love that has nourished my higher self. The range between selves is so vast now. I thought I'd reunited into wholeness, but maybe that was just a moment. Or perhaps *this* is.

Thoughts on this "holiday":

The kinds of people who celebrate colonialism have likely never considered that there is any other side of it. There are flags out today, suggesting we are proud. It is an elitist and horrifying brand of patriotism.

Columbus is only one of innumerable historical examples that we, as children, are told to worship as valiant explorers and conquerors: history is taught to us from this singular, white perspective. The truth about these men is always some version of a centuries-long tale of First Peoples being used, murdered, enslaved, suppressed, displaced, and disenfranchised, if not erased completely. To this day, we still collectively minimize or disregard the history and culture of anyone outside the Eurocentric patriarchal exemplar, especially those that model has deemed dispensable in its rise to governance. The injustices against Indigenous people became woven into the fabric of this and so many other nations, starting with the conquests of men like Columbus.

The agents of imperialism are increasingly more encoded and furtive now, more embedded in our value system based on wealth accumulation, though the outright violences remain easy to find as well. What the American collective doesn't say out loud, it says with our laws steeped in racism, our systems designed to exclude, and our continued eagerness to take what isn't ours. In the white rise to economic power, there has been a thinly-veiled rebranding of the systemic injustice against Indigenous people and People of Color, a shift from openly committing atrocities to the creation of a paradigm where they are silently built in.

The new colonialism is less blatantly racist, and more obviously classist. Those who have been stolen from, devalued, and oppressed for hundreds of years have had no opportunity for a voice within the system, no equal footing for economic or societal standing by design. And those in power who vie for further financial gain are culturally excused from moral responsibility in the exploitation of resources or other human beings in the name of capitalism. Our national loyalty to individual gain, our pervasive religious sense of dominion, and our childhood educational indoctrination provide the platform for a society more than willing to shroud its white guilt and abhorrent dislike of otherness with its love for material accumulation.

There is a whiteness far more insidious than that which openly bestows pale-skin privilege. There is a cultural whiteness that devours groups of people into its amalgamation and spreads forth into the reaches of the world, consuming and conquering, horrifyingly out of balance with nature and incapable of right relationship. It propels everything it touches into unsustainable greed, leaving a wake of collateral carnage in its bloodlust. We are either complicit or fighting against this- it is a virus that thrives in the numbing lull of the machination that keeps privileged classes ignorant or indifferent to its vileness. We continue the new colonization like its our birthright, propped up on the false heroism of men like Columbus. We take whatever we desire globally, either by our outright sense of entitlement or by installing puppet governments under the guise of liberation. We whitewash and Westernize the world over, spreading our linear and unsustainable values, overprinting Indigenous ones. We corrupt with money. We oppress socially, politically, and financially, then we shrug and point to the system if someone demands culpability. In our own country, having already largely stolen every available material resource from Native people, we continue to take from them in subversive ways, arrogantly appropriating sacred ideologies and commoditizing symbols, suppressing their autonomy with eminent domain, and turning away assistance and resources in times of need.

There has never been compensation for the stolen lands and stolen lives, for the lost cultures and lost dreams, there has never been atonement for the broken backs of those upon which this wealthy nation was built. White privilege, rich privilege, is so indoctrinated that people lash out, aghast like *they* are being attacked when it's exposed. The oppression continues masked or unmasked, and severed from historical reality or even a shred of white accountability. The collective says: it's your fault if you can't succeed here.

It's reprehensible.

Many with privilege live comfortably ignorant lives, never understanding the system from which they benefit. Money is the modern separator of people from inclusion, and our national and generational wealth was created on atrocities. While there is a tendency to blame the poor for being poor in our capitalism paradigm, there is a racial parallel that blames non-white people for being poor... even though it was with stolen resources and forced labor that affluence was generated for their oppressors.

There is an obligation for people randomly born into advantage to step into the discomfort of our history and question our current contributions to the world. There is a dire responsibility for all of us to awaken to- and halt- the relentless spread of a value system that imposes itself violently upon nature and other people. It is up to us to seek to restore relationship, to ask for access to the lost histories we were never taught, to vote with our time, energy, and money for the future we want. Only if we are awakened into deliberation and clarity can we begin to reverse the course. We must educate ourselves on how we can best be personally and collectively inclusive and accountable, and how we can be good allies. It is up to us to protest holidays that exalt murderers.

Globalization is destroying that which makes places unique. The frontiers of empire-building are no longer physical endeavors with ships crossing oceans, they are the slow but steady corporate erasure or homogenization of ways of life. The new missionaries spread gospel about the religion of money not just by creating demand for new things, but by creating wealth disparity that cannot be survived without playing the game. As places become richer, ethnic diversity is flattened. Marginalized groups are priced out, kicked out, or sold out in the evolution of economic "progress". When places lose their Indigenous people, their diversity in general, they lose with those groups an immeasurable social range as the art, language, music, food, architecture, events, traditions, and ideas are diluted, absorbed, or commoditized.

In the US right now there is an increasing fear of interpersonal otherness that runs tangent to the corporatization of commerce. The underlying racism within the whole system is rearing its head.

What would there be left to learn if places or people were all the same? Sameness dampens our holy curiosity, mutes our ingenuity, stifles our ability to thrive and adapt. Our differences are what make each place in the world special, each soul in the world necessary, each perspective worthwhile and critical. The global advancement of resource-based economic systems promotes not only unsustainable population expansion, but the deterioration of the cultural range of humanity.

As communities in the developing world raise their standard of living, they often sacrifice their connectivity with each other and the land. The way they spend their time changes in order to rise from subsistence living to drawing a paycheck. It is immensely complex from the outside to understand whether or not people *want* this change, whether they have a choice, and whether they can foresee the opportunity cost.

In my travels, I've noticed that people in very poor places often seem *much happier* than people in the US, something that diehard capitalists would rather not believe or address. But it makes sense to me... they are not happy for being poor, they are happy for being *connected*. It is hard to see from within our culture of accumulation and vanity that there is an inverse relationship between consumerism and connection. In many places I have now seen, they have not traded family time for a commute and a job, they have not traded creating crafts for mindless time in front of a TV. They have not irreparably damaged the land to produce an excess to sell it for money. They tend, more than in developed places, to live intimately with each other and their environment.

This way of life is dying but I am running toward it with open arms, desperately craving this belonging, seeking some vague admission back into the Real World where the right things matter.

You will have many mothers
And none of them will be
the portal through which you arrived in the world.

And this is ok.

You were born the daughter of a vast humanity
Created through a long lineage of souls connected
Held by the Everything,
Not to be kept by a person.

Trust that you are held
Even if there are no arms around you.
You belong to the universe
Swaddled safely in the network of love.
Trust that you are loved
Even if there is no one around to remind you.

Other mothers will feed you
And you will learn to feed yourself
You are no orphan, child-
You were born from stars.
Don't despair for your lack
But rejoice that you are here
Awake,
Awake
And full of light.

There is a drive toward synthesis of a narrative for my pain, as though if I could sufficiently flesh out the story of it, I could ask it where it still hurts. The work of forgiveness and softening is most potent when it has a specific place to be applied.

I am creating a map of this inner landscape and its defining peaks and cliffs and faultlines. I am traveling to visit the harshest terrains, giving these places names, and sitting with them until I am delivered into their wisdoms.

I've been re-parenting myself now at age 36. It's a little late but I still have within me the child that needs someone to do this, and the only person it can be is me.

I came face to face with her again, after I'd abandoned her over and over and tried to forcefully make a life without her. A grown up life, I decided somewhere along the way, doesn't have reasons to honor a small, ashamed internal voice. But she looked for a parent in abusive lovers. She tried to feel worthy by living out the ways other people told her equated to success. She couldn't manage her emotions so she buried them and numbed them until they turned into mood disorders and autoimmune disease and alcoholism.

I spent years shutting down this child, showing her she didn't get to make my choices for me; and all the while the mistakes I was making were a reflection of her pain, her lack of comprehension of gentleness, or belonging, or acceptance. Even though I wouldn't look at her, time after time she showed up asking me to understand as my life materialized the unresolved injustices committed against her.

How can we help what social imprint we internalize when we are small? No one escapes the formative years without being wired in a way that will eventually need repair. I saw that obedience meant approval, doing what was expected of me without asking for anything was the best way to get positive attention, that everyone had problems so it was best not to burden anyone with mine, that respect was allocated like a weekly allowance. I saw that happiness was the only acceptable emotion to show, while others were flaws meant to be fixed. I saw that love meant sacrificing for people without necessarily connecting deeply with them.

In my formative years this conditioning, saddled with an unnamed emptiness from what was really lacking, led me into childish dead-end relationships. I leveraged sex for the approval and comforts that I thought were love. I didn't have the luxury of healthy relationships modeled for me to see the difference. I didn't know how to ask others for what I needed because it had been programmed out of me to ask myself what I needed- or to feel important enough to need at all.

I fumbled along trying to satisfy others, both romantically and professionally, with the shrouded core belief that if I sacrificed to please them, they would give me attention, approval, and support... and I would therefore feel loved and successful.

Except I didn't.

I just felt more empty.

I know now that I used this people-pleasing means of social currency for at least a decade of my adult life. I couldn't understand why partners left me, why material measures of success gave me only fleeting moments of happiness. I felt doomed to be lonely no matter who I was with. I fell in love with people who might save me from myself, with little regard to who they were in real life. I felt chained to a prescribed educational and career trajectory, with my milestones of success- graduations, jobs, promotions- holding only as much meaning to me as the approval they gained me with the father whose dogma I was living.

The universe has a way of asking us to listen. It's subtle at first, a whisper, a gut feeling... but if you ignore it long enough, life will inevitably arrive at pinnacle moments that are a culmination of many choices made while not listening. There, in those moments, we are tested. The universe is screaming: *NO! Change this! You are out of balance with yourself and others, you are out of touch with your truth!* Whether or not we pass will determine if we level up in consciousness- or develop a protective cognitive dissonance to pacify our subconscious discord with our own emotional and developmental limitation.

My inner child, that tiny, wounded, awkward goddess whose

maladjustments and misunderstandings poisoned my life despite my ignorance to this, finally made herself known to me in the throes of a tragically codependent and toxic relationship. The whole relationship was two years long, but by the second year I had become a shell of myself: defensive, withdrawn, indecisive, secretive, and desperately depressed. The break up itself was four months long, a teeter totter of manipulation, renewed hope contrasting violent words, blackness, and despair. I wanted to evaporate, dissolve, to cease to exist. I wanted to die. My identity had been crippled as I had become what this person projected onto me.

But as that was happening, something else was also taking place. Undergoing the death of everything that I had become to please someone else had gotten so literal that it sparked the teeniest, most private, secret fire of defiance somewhere deep inside of me.

And even in the months of turmoil that became my life through the ending of this relationship, and the gripping depression that followed... that spark grew bolder and stronger, and eventually pulled me out of the darkness. Some part of me, some primal, forgotten self, wanted to heal and survive and thrive more than I ever had before. That spark would not be smothered by another or ignored by me any longer. It had taken almost dying to *really* want to live. It had taken the hardest relationship of my life to understand love. It had taken someone who treated me worse than I treated myself to make me finally protect my Self. And through the portal of this extreme pain, I was born into love and clarity again.

The healing has come in two parts, *that* I can see since I'm past the first part. Recovery from that specific relationship was the first wave. It was the hardest work I have ever done, as no wound has ever been so deep, so severe, so open for the world to watch bleed. I had to reconstruct my reality and my identity. It's so finite, so packaged, to write that sentence, but it was vast, impossible, cosmic. I surrendered to gravity to pull my pieces back together because they had no will of their own. At the time I didn't think of this gravity or believe in it or trust it, I only knew my capitulation. I could have just as easily disintegrated, but that concept had no meaning to me then, nor did any other. In my raw and open state, I entirely gave myself over to whatever would come. I sat with my pain and I tended it, I gave it time to be, because there was nothing else I *could* do. There was no way to numb it this time, there was no energy left to resist it. So I sat. And slowly I recrystallized. Only after coming back together did I recognize that the Universe held me, and that there was such a thing as the divine law of emotional gravity that acts on us when we are shattered apart.

It is easy now to inadvertently gloss over how graceless and ugly that process was at times in my daily life. Healing came in stops and starts, mistakes and breakthroughs. There was no womb or cocoon, no container for the undone self, no hiding until I was more finished. I couldn't shelter anyone around me from the hurricane that gathered momentum as it consumed me and reworked me in its image. I was a rag doll, I was a weapon, I was a natural disaster. I went to work like this. I traveled like this. Made new friends and took new lovers like this... open and unarmored and bouncing off of life with my enormous and hastily packed baggage. There is no pause button for life, nothing slows down for you to be reborn.

The second phase of healing was born by the first- I am finally at the point where I'm listening. *What do you have to tell me, Universe? How can I restore you to holy belonging, child? How can I create a Self out of truth and love so that my path will be full of integrity? How can I honor my time in the world by being deliberate with what I choose to do? How can I heal the parts of me that feel numb, unloved, or unlovable? How can I nurture my inner child into wholeness so that she can seek right relationship?*

It is here where the re-parenting is happening. Here, I am rebuilding myself without the weight of maladjusted coping mechanisms, uncovering those one at a time and resolving the reasons I created them. I'm slowly but absolutely replacing every one of the cells in my body with ones that I create out of love- the love that comes from that spark that has now turned into a blast furnace inside my heart.

For every skinned knee on that tender-hearted child, I'm showing up now to put the kisses and band-aids. For every time she felt abandoned by someone else, I'm showing up now to tell her that I'll never leave. It was *me* all along, I tell her. The one you looked for in others: the approval you needed was your own, the respect you deserved you were owed from yourself, the attention you desired was that of your own heart, the love you craved so intensely was the love you have right here to offer yourself.

I know that if I give myself these gifts, like a loving parent to my own little inner child, that I can raise myself into the person I want to be. A person full of light and integrity and divine intuition. If I show up for myself I will never be lonely when I'm alone. If I have internal clarity and follow my heart, it will lead me right where I am supposed to be. It's my highest work, my greatest art. It's my most important relationship and my most inspired education.

So what do I want to be when I grow up? *Love.* And I'm practicing right now.

Oaxaca and Chiapas

Returning to the bigger river,
The one where I was born,
Spawning.

The journey up and into the heart of things
to fulfill a destiny not yet revealed

Freefall into the waiting arms of the universe,
The open net of a broad but familiar humanity
Oh, how I've missed being held by you

I don't know where I am going
But I know I am exactly where I'm supposed to be.

Already I can tell that I'm different, and maybe this place is different too. Can you ever *really* return to a place? Will you fall in love the same way again or will you overprint your memories, replacing your daydreams with disappointment? Some things are meant to be moments in time and some things transcend. How can you know unless you go back? My Spanish has improved greatly since I was here last, but in some ways I feel more like a stranger. Knowing what is going on seems somehow more frustrating than fumbling happily through a beautiful mystery.

The first day is always the most challenging, most opening: acclimating to the vibrance, the colors and street noise, the rolling cadence of overheard conversations like a song playing in the background. The buzz of the upcoming holiday permeates the place, there is an electricity in the air. The walk to breakfast softens me bit by bit into relaxation within the pace here, preparing me for the markets and madness and faster or more confounding experiences that will follow. It takes time for me to feel at ease with my lack of understanding, with the degree to which I clearly stand out. It's best to take the day in parts, with an intermission to regroup and process at the hostel.

I choose a small table by an open window on the second floor. The colorful papel picado banners flap in the breeze above the cobblestone street below me. I load my coffee with sugar and let my mind drift.

The peace of my thoughts is interrupted by a group of men who join me in the otherwise empty restaurant, and I am immersed in the harsher aspects of the language, dominant and masculine, spiked with laughter and all the curse words I know. It's in these situations that my curiosity slips toward uneasiness. It's clear they are referencing me in their conversations, but I don't understand enough to know the context. I pick up that they don't think I can understand them at all. I pretend not to. My intuition tells me that they aren't dangerous, just lascivious.

As the disquiet of their imposed disruption starts to twist, I decide I have to let go- let go of needing to know, or caring how I am perceived, or involving myself energetically in an exchange with them. The way they treat the lady serving them is demanding, and I wonder if it's an aspect of the machismo culture I'd never noticed before, or if they are just rude. Men in groups have a familiar dynamic anywhere I've ever traveled.

I gaze out the window, going back to life in my bubble. Their proximity and loudness changes the nature of my experience but it need not affect my self-consciousness. I buttress my tranquility. My hypervigilance is no doubt a product of the jet lag and exhaustion.

Suddenly they rise, throwing some pesos on the table. They bid me *buen provecho* with brazenly charming smiles, then they leave as abruptly as they barged in. I'm left wondering if they were really as crass as I imagined or if I am just in my head. I decide that non-judgment needs to be more at the forefront of my mental process as I move through the day. Hopefully I can get myself further below the radar soon.

I will go on more tours here this year. I'm trying to keep my expectations in check since last year overflowed my heart. The cemetery visits absolutely changed something inside of me, it repaired some sense of community and family that I didn't know I could feel a part of, especially as a tourist and an outsider. The societal perception of death here is at once reverential yet... messy, human, celebratory, complex. It was Everything.

I found my hand-scribbled notes from last year.

Dia de los Muertos cemetery tours, Part 1:
I find the energy of Day of the Dead very confusing. Part carnival, part Halloween, part funeral. You see at once kids laughing, people dressed as devils charging to take photos, and people sitting solemnly with elaborately adorned headstones. Smells of copal, marigolds, lilies, and food fills the air. Tuba-laden brass bands pound out minor key funeral marches and dissonant polkas that remind me of Tim Burton movies. Ice cream, glow in the dark devil horns, and cigarettes are for sale on carts. The tourists add a voyeuristic element, but the guide insists that the people with loved ones buried here do not feel intruded upon. I somehow feel exploited for them, perhaps projecting my cultural heritage or just sensing an underlying disrespect in the hoards of people treating this as a spectacle. If one doesn't have dead here, it could just as well be a strange music concert or an interesting place to drink with friends. The living and the dead, side by side, in such an uncomfortably familial way... maybe this is the reality of it all. What's missing for me is the feeling of loss or grieving. I might be noticing a cultural difference or those in mourning may be guarding their emotions since it's such a public affair. I'm the kind of person that always tries to get pictures without people in them, but to cut the people out of the shots here is to miss the point entirely. The people are already missing. The grave stones are memento mori, people without people, disembodied lives lost to history. The living press on in the contrast of continuance, ridiculously keeping to the surface of our emotional experience, or conversely and more

positively, despite our pain and losses, continuing boldly to live.

<u>Dia de los Muertos cemetery tours, Part 2:</u>
The last cemetery we visited tonight brought this custom home for me. I internalized more here than in the larger cemeteries. Santa Maria Atzompa was different, smaller, more intimate. It was when I experienced this place that I understood that the others had been Real too, but that depth was veiled by the layers of tourists and vendors and the recent enmeshment with Halloween.

We were given candles and flowers to put on headstones of our choosing. People here seemed more emotional, more openly vulnerable. The walls were down. There was pain. There was longing. The dead would be lead back by candles and water and mezcal and marigolds, but things could never be the same as when they were living here, among those who loved them.

Candles and flowers and incense and music- quieter here, asking. Thick, cool air and a palpable waiting for the return of lost family and loves. Anticipation and ache.

I shared my candle with a lady who sat solemnly by a sparsely decorated fresh mound of dirt. I gave half my flowers to a grave with none, and half to the grave of a child. At one point, my eyes met those of a man sitting alone wrapped in a blanket by a headstone. My hand went to my heart and he nodded, his eyes filling. I sat with him, held his hands, and cried with him. We never said anything to each other.

What a difference a year makes. The last time I was coming here, I sat in the Phoenix airport at the gate, terrified that you might actually come. You hadn't spoken to me in weeks, yet you still had your flight ticket- we'd bought them together. I was in knots thinking that you'd show up, and wondering what that would mean for us, for me, for this trip that had long been on my list. I knew that if you didn't get on that plane that you were cutting the final tenuous threads that still bound us.

I came here alone that day.

And to the surprise and relief of my tender, bruised heart, the warm arms of humanity wrapped me up and held me. It was here that I believed for the first time that there was life after you. This place, the people that I met, the way I felt here, the entirety of my experience snuck into the trembling fortress of my ruined self and opened the gates again. With every passing moment, I became increasingly glad that you had chosen not to come. I let go and let the river of life, the pulse of this place, the smiles of strangers carry me. I did the things you'd never do: I made new friends and had conversations that ran into the hours of daylight; I ate at the best restaurants and ordered without thoughts of budgets or time or plans; I drank mezcal with families that were gathered in cemeteries, reflecting with them on life and death and love; I hung out with locals, I went to their houses; I went where I wanted; I laughed and loved, in the flow of whatever came.

It was the prying open of my soul that your twisted love and control had nailed shut, and it was the first step that I took into a new world, one without any trace of you. One that I'd never show you or talk about with you, one that I'd never offer you to judge or poison or take from me. And it was the most cherished, beautiful gift.

Returning here I remember those early moments of opening, and I love them for what they were. The quality of my openness has changed in the last year of my life. It's become less novel, and considerably less reckless. *These things are good things*, I remind myself. I have healed so many of the wounds you gave me, and even the wounds that were there before you, since I first came here. I have learned to feel when my heart is closing and gently support it back into bloom the very instant the petals retract. I am learning to no longer participate in relationships that compromise my self-worth or leave me feeling knotted up in hopelessness.

There is no longer the immense rush of relief from freeing myself or surrendering to my life, it is now my equilibrium. There is no longer the complicated, smothering toxicity in my life from which I needed to escape. But I will always remember with gratitude the way this place caught my freefall in the very way that could restore me: with love and light and an explosion of sensory excesses that made me feel radiantly alive again for the first time in a very long time.

Thank you for not coming here, thank you for letting me go.

On the flight from LA to Oaxaca, my seatmate was a petite woman with kind eyes who nervously wrung her hands together as we prepared for take-off. She hated flying, she told me quietly, laughing at her own dramatic reactions as we rattled along through the afternoon cumulus.

We fell easily into conversation. I think it soothed her nerves to have a distraction. For me, I couldn't help but to instantly adore maternal figures when they arrived into my life.

She told me that she was from a small village outside Oaxaca and was coming back to see her ailing mother. She spent 23 years in LA and raised her kids there, but Mexico would always be home. Last year, her 20 year old daughter was one of three women nominated to be Queen of their village. The families of the chosen women work for a year to raise all the money they can. Then all the funds are combined in a ceremony, and every penny of it is spent on an elaborate four-day party where the whole town is invited to eat, drink, and dance- all for free. The Queen is chosen by the amount of money her family raises for the party. This is their tradition of Carnivale, and it happens in February.

She told me that her family had sold everything from Oaxacan cheese to handbags they imported, in addition to working their regular jobs. She took no days off and barely slept. Every dollar that they made went to support her girl's nomination. They raised $70,000, and her daughter won! She was astonished at her own feat even as she told me. It's an enormous amount of money for anyone, let alone the people in a tiny, remote village. And to use it all on the extravagance of a party! But while she explained her embarrassment at the indulgence she also glowed with pride, and I sat in awe of the lengths of sacrifice that people of this culture will go to for their community and traditions, for their families.

It's no wonder that it's one of my favorite places I've ever been.

The last time I traveled here, I felt so thoroughly emotionally alone that I couldn't have noticed my physical aloneness except to revel in it. Now that I feel more connected with the people in my daily life, there is an occasional twinge that edges into my solitary experiences.

Dinner at a table for one triggers my heart into asking me to tend it. So I respond internally, *I'm right here, I'm listening, we're having dinner together*. And weirdly, instantly I feel happy again. It really just takes this acknowledgement to restore my mindset to some innate calm. It's new for me to feel the finetoothed comb of introspection pull gently through the knots of an experience without diverting to self-judgment or self-medication.

There is a fine art to dining unaccompanied, to taking yourself on trips to explore the world, to being good company for yourself. I am still learning, but it's an empowering and enlightening endeavor. I am breaking the bonds of codependence, I am getting to know who I am and what I like. It's worthwhile examining whether or not you are interesting company for yourself, and traveling solo provides both the time to examine this and a wealth of odd experiences to enrich a personality, I think.

I turn off my phone, committed to being in the moment. I remind myself that this is a practice, permitting my attention to reside in my own companionship without seeking diversion. Being alone can still be social. I am out, enjoying the evening. I am present in my own experience and in that of all who are around me. I am present in my connectedness to humanity. I am at the most idyllic rooftop restaurant watching fireworks erupt in the warm night air, and I am simultaneously very alone and not alone at all. This is the strange paradox of what it feels like to be alive now.

The new colonization is has its cold, pale hands on the shoulders of this place. Quietly, the advancing front, a slow poison of English-speaking retiree expats, have ridden in on a wave of tourism dollars that support mega buses and American restaurants. No wars are fought as the bland, convenient, commercial culture is superimposed on the Indigenous and post-conquest local identity. The essence of places are always in flux, I realize, but I feel especially sensitive to the ways of modern corporate erasure.

I am a tourist here, but I sit saddled uncomfortably with my loathing for the tacky tourist armature, a flimsy plastic mask haphazardly tacked onto grand and ancient monuments. The backpackers are mostly too cheap to affect the economy much, and collectively seem enthusiastic about integrating... but the Boomers, who are affluent enough to impact the expression on the face of franchise leave a deep mark. From what I have witnessed, this subset largely seems less interested in learning the language or assimilating into the existing lifestyle, and more prone to residing stubbornly within their indoctrinated culture, no matter where they happen to be in the world. Wherever this demographic swells will nearly always be effectually spoiled with box stores and chain restaurants soon to follow, with prices inflating and further tourism becoming necessary to prop up the expansion.

I wonder if this is *progress*, simultaneously lifting local people out of poverty while eroding the distinction in their towns. *Can we do better?* Are there ways to implement direction for this growth? Can we somehow protect places from melting into consumerist ambiguity as they become gentrified, Americanized snowbird destinations for people with disposable income? How can I make sure I am not a perpetrator of this wave of influence that I despise? Am I already a part of the problem? How can I both visit and protect a place?

When we find ourselves in judgment it's a good time to reflect on why we are separating ourselves... what are we afraid to see or admit? Am I just as guilty as anyone else who has come to enjoy this city?

How have I commoditized or fetishized cultures that I've fallen in love with? If I am going to travel, how can my influence benefit rather than destroy the very places that I love?

The irony is not lost on me that I write this from a rooftop bar drinking mezcaltinis. Tourist infrastructure has no doubt created my current environment. Maybe *my* privilege is the source of my discomfort, the thorn of my own cognitive dissonance the origin for this thought.

Avoid eye contact
When you see that I'm old
I will now perform for you
The dance of the dying eggs.

We are not separate.
To isolate creates a wound
That severs us from our Selves,
From our fellow people,
From the dead.
Understand this space
With your touch-
Put your fingers in the dirt of the earth,
In the dirt of the graves.
Kiss the hands of the sick and weak,
Caress the scars that decorate your heart,
Push the hair back from the tearstained cheeks
of your inner child.

The cemetery is full of life
Your open wounds are full of medicine
Because it is within your blood.
Go back inside,
Feel your own inclusion.
Every time you touch the separation
You close the circle into wholeness.

What happened to me last night? Was that a dream? Real life? Something supernatural? I dreamt (?) that I was in bed at my hostel- this same bed, having a seizure or some issue that made me convulse. I was thrashing around, in and out of consciousness, seeing and hearing only in clips. I heard voices from the terrace saying, *Who is making that noise?* In a more lucid moment, I was embarrassed that I had been screaming in my dream and had woken other people in the hostel. I couldn't breathe or see correctly. Then there was a man in my room standing over me, and I was panicked that he would hurt me. I didn't understand what was going on. Was I sick? Was he going to rape me? Take me to a hospital? Was this real? Everything was blurry and shaky, and I couldn't keep my eyes open or talk. My heart kept jumping like it was being shocked.

When I "woke up" I was alone in my room. I could not tell if the man had just left and I'd recovered from a seizure, or if I had dreamt the whole thing. I checked the door and it was locked. I asked Grandma to protect me, and I went back to sleep. I want to believe it was a dream but I still don't really know. Was I two nightmares deep? Did some part of that really happen? Was I spiritually attacked by something? I feel oddly embarrassed at my hostel as though at least the screaming happened in real life.

The veil between worlds here is so thin right now, and I need protection. That was among the most bizarre things that have ever happened to me.

We left the festive insanity of San Agustin Etla and the increasingly wild, costumed comparsas, the four of us new friends chatting happily with the rest of our fellow tour-mates. Whatever lie ahead that evening, we were entrained.

The van bumped down a dusty two-track to a tiny village where the annual tradition on this night is to dress in costume and go house to house putting on a play. The script is about a man who dies- doctors try but can't save him. His widow calls in a witch to raise him from the dead. Once he's reanimated, the brass band starts up and the whole village, every single person, dances triumphantly to a few upbeat songs. Everyone drinks a "mezcalito" poured from a repurposed Coca Cola 2-liter, then they take the parade onward to the next house to repeat the act. Over and over they perform the play with increasing hilarity, collecting the people from each house into the procession along the way. Apparently they sprinkle into the play all the funny things that have happened in the village throughout the year. We were the only tourists. The magic was thick; the buzz of connection, overwhelming.

We walked through a cornfield in the black night and crossed a small footbridge over a river. Awestruck, we watched the play again and again, wandering the dirt paths between homes shoulder to shoulder with the villagers. We *cheers*-ed our little plastic cups of mezcal, and they were ceremonially refilled by a beaming middle-aged lady. We flashed knowing smiles at each other as we spun along, dancing exuberantly with locals dressed as witches and doctors, devils and angels. Passing hand to extended hand through the crowd, wrapped up in the utter excitement, we allowed ourselves the unfamiliar feeling of tolerating deep inclusion. Strangers almost instantly became family.

It's amplified when you're traveling alone, and you have to *let all the way go* to feel it: you can not *remain* alone in the world for long, we are connected. We are designed to make friends, to trust, to love. It's the reason I keep jumping off of new cliffs... I ache to feel that net wrapped around me again when

I forget. Each time I trust a little bit more that it waits for me, that I will meet kindreds, that even if we speak different languages and come from vastly different traditions, there is a baseline, a humanness that unites us. I want to feel the togetherness, I want to touch each other, I want to live on the plane of reality where we are one family, and we are family with the world around us.

"It's the most I've ever let go of control," James said simply. And I loved that reflection on the evening. Who could know where we were, or whether phones worked here, or what would happen next? Who could care? On that night we were all related and elated, present... and my heart overflowed like homemade mezcal sloshed into thimble cups while dancing.

My heart fell out when I was a child
And I stood there
holding it in small hands
Unsure of what to do
So we put it in a glass of milk
to keep it good
until we could put it back in

Or maybe that was my teeth

Crumbling like a loss of control from a hole in my chest
Bloodied mouth
And
The days of not being able to love until
My new heart
My grown-up heart
came in

My heart got broken so I put it
Under my pillow
In the bed we used to share

Or maybe it was my childhood bunkbed

But I didn't wait til morning to check for
Compensation
Or a token to make me believe in magic-
I put it there to hide it where you'd never look

Gnawing ache in my chest
Ravenous futile gnashing against
The void of a barrel cage
Dry socket
That waits to be filled anew.

Today I went on a tour to the small Indigenous towns of Zinacantan and Chamula in Chiapas. I was the only native English speaker, but since the guide's second language was Spanish as well, she spoke clearly and slowly enough for me to follow. Marina was born in Zinacantan, but she wanted more for herself than life could offer there. When we stopped the van at the edge of the community, Marina told us the story of her life. The levies of her eyes spilled over when she explained that although her family still lives here, they have disowned her for getting an education and not following the traditional ways.

In these village cultures, women are literally bought and sold. Men can have multiple wives, and they all live together. If the women give birth to girls instead of boys, they are considered worthless. Girls marry when they turn 12, and they do not get to choose. Women are not allowed to voice opinions. The villages do not value learning outside the home, and there are no schools. Marina wanted an education and a better life so she left her family of 25 brothers and sisters and went to San Cristobal when she was 12. She had no Spanish, no English, and had to sleep in the streets until she could make money. She fought her way up and went to university for five years. She never married because the men here do not appreciate her independence- those that became interested in her would end up making her subservient, something she could not tolerate. I loved her immediately.

She led us inside a concrete building to see a small traditional altar- one that combines Catholic influences with Mayan. The floor was sprinkled with pine needles for fertility. Various figures of saints, lit candles, crosses, and flowers decorated the space. Strands of rope garland were tacked to the ceiling, and globe-style Christmas ornaments hung in patterns. Marina explained that those aspects were not Christian, they represented the universe with the stars and planets which were very important in Mayan Tzotzil life.

In the next room we tried on traditional clothing, and then we were led back into a small adobe-and-tin hut where a young woman made us fresh tortillas. We tried typical foods, consisting of beans, seeds, and pungent cheeses. The woman then brought us mugs of traditional coffee, which had other other types of seeds and sugars in it.

When we had finished, we piled back into the van. At the next stop, we trekked up the steep hill where shamans come to take their special plants and pray to the sun, moon, and rain.

The last stop, a few miles away, was the village of Chamula. We drove through the winding hills covered in small squares planted in various textures and shades of chartreuse, celadon, and rich forest greens. The wildflowers popped like splashes of sunshine, and the corn stood ready for harvest in crispy golden browns. All the plots were alive with people hand-tending them. The patterns and vibrancy immediately brought to mind the creative textiles from here, and it was easy to see the references they wove. We drove past women in traditional dress washing clothing in a small stream, their ankle length heavy black fur skirts dripping from the work. On their heads they wore woven scarves in a folded-flat way, their hair in long braids.

We arrived in Chamula and were warned that in this village we must not take pictures of people, nor the inside of the church as it was very offensive. The sun was intense here, the air was thick with wood smoke and the music of drums.

We walked through a narrow street where people were gathered around small fires made on the concrete, over which boiled giant pots of whole chickens. Some women stirred the pots, some crouched or sat in the shade nearby. A few children played. Men mostly sat separate from the women, a reflection of their social distinction. The ground was littered with sticks, feathers, bones, random chicken parts, plastic trash. Marina explained that the people gathered here and ate together on Sundays and when there was a party. Six people squatted around one of the big pots of cooked chicken, reaching in communally, serving and eating with their hands.

The white and green chapel of San Juan Chamula stood unassuming at the edge of a small open market. Walking through the wooden double doors into the dark space, my eyes took a minute to adjust. The inside was a dim and gaping cathedral with an assemblage of hundreds of lit candles in colors representing the sun, moon, underworld, rain, and mountains. The candles lined simple tables and floor spaces around the altars for saints along each wall, adding to the spent wax monuments of their predecessors. The floor was entirely covered in pine needles. The dark ceiling was painted with jaguars and stars. Blue smoke danced in the geometric arms of light that reached down from high windows.

Throughout the church, groups of people gathered together, having cleared spaces to pour wax on the floor and put rows of candles in symbolic arrangements. Their private rituals incorporated live chickens, eggs, orange soda, and posh (the corn and sugar alcohol that is sacred here). They squatted, swaying trancelike, chanting softly or spitting posh at the candles to make them flare. Marina explained quietly that the villagers did not believe in doctors and very frequently died at a young age. Shamans came to the church to practice ceremonial medicine, curing people by rubbing a live chicken over their body to "absorb" the illness, then sacrificing it. The people would later eat it with their families to complete the ritual. And also, I assumed, not to waste the meat.

Missed connections.

Sometimes the disconnect is that I still don't know how to be loved. Sometimes it's that I don't know how to love myself, or that I lose myself in the act of loving another, or that I have nothing to offer someone when they arrive. Sometimes it's that someone else doesn't know how to love to me.

Maybe at times we come into each other's lives needing something, looking for something, so desperate to find the missing piece that we are willing to see it anywhere. Sometimes we use each other, whether intentionally or subconsciously, making another into a savior, an escape, a character in our pre-written script, a manifestation of our fears or desires.

And maybe this is happening to all of us at once, we each always have something we are healing from or running toward. And we spend our lives bumping into each other, ricocheting off of our image that we find in other people, rejecting those that trigger our *samskaras*, clinging to those who taste like our birthday cake when we were five. We attach like magnets to people who aren't right for us, simply because we are designed to attach. We repel people who would be lovely company for us, because we happen to be upside down in that moment.

Love is timing, even between the right people. It is a moving target not just because of the patterns of others in their own orbits, but because of our own cycles of growth, our accumulation or unpacking of baggage, our breakthroughs in clarity.

Loving yourself, for those like me, is a work of art still being created. It's a kaleidoscope being twisted toward the moon: a beautiful, painful dance of collisions and recoveries, occlusions and revelations.

I am binaural, vibrating in two frequencies that run parallel and do not touch or interfere. I am at once high and low, light and dark, a sunshine seeker of high truth and an indulgent shadow self. It feels more intrinsically human than pathological, I am growing big enough to hold the separate selves as facets of the same reality. I am not split apart, I am universes converging. I am music, with its parts commingling within the senses of the observer. I am of this Earth and of this Time, I am their daughter, their spitting image. I feel the potentials concurrently within me, I feel the range, the effects of my animal chemistry, the edge of consciousness, the choice.

Dream:

I found a really big caterpillar- it was rare and lovely. I told Kimmy not to pinch it or pick it up around the sides because it breaks their teeth. Then I saw another butterfly of the same kind nearby. I let the caterpillar crawl on my hand, and I put them together. The caterpillar squeaked. The butterfly turned into a huge rabbit who grabbed and ate the caterpillar while lying on its back like an otter. Another butterfly nearby was dead and one had wings that caught on fire so I blew them out.

This dream is a warning. There is a heavy undertow here and my dreams dredge these things up so that I can protect myself. I am in the process of transformation, and it is dangerous to make assumptions about who is like me. I am aware of how to protect myself in one way (agency), but ignorant in another (sexuality). A naïve part of me is in danger from something that is not what it seems, most probably a man that I am attracted to.

Seeing him out of my bus window seized the air in my lungs. In a city of a quarter of a million people, there he was, mysteriously and randomly in my life again. I wasn't prepared to confront him or even to ever talk to him again, but the moment I saw him boarding the same bus, I knew it was in store. I must have subconsciously suspected this was possible. Maybe I ended up here for exactly this reason, on a collision course with an overdue resolution, whatever that might mean.

The intersection of our lives had been short-lived, by the fault of circumstance or the fates, the curtain was being drawn no sooner than it was raised. Still, it was a sincerely written single-act play, a brief but lovely convergence of worlds. I attributed a quantum leap in my healing to our time together in Oaxaca last year. It was a somatic reprocessing of connection, it was proof of life, a renewal of potentials. The final scene would reveal the fleeting script as a minor tragedy, with petulance punctuating a wake of realization that the encounter was intransient. Although we were never slated for longevity, it stung to receive a message months after my departure declaring it in cold absolutes. I wasn't one to harbor judgment for an impassioned reaction, but it would be our final correspondence, the fade to black.

A year to the day after meeting him, here I was again, staring stupidly as everyone settled into the blue cloth seats. The folding door thudded closed, sealing the twelve of us into the container of an experience.

He didn't see me until we arrived at the village and took our places at the long wooden tables. His jaw dropped when his eyes landed on me. I shrugged sheepishly, resigned to the awkwardness. We hugged as old friends, and the anxiety of an ominous public spectacle dissipated, yielding then to a wave of residual questions. The warmth of the reunion made it even more strange that the end of our communications had been so terse. What had happened? I was interested enough to ask but detached enough not to press. He said that he'd explain everything, that I should give him my number so we could make sure not to lose each other in the crowd.

And so I did. Maybe out of morbid curiosity and maybe because I thought there was something there I had missed, something to be forgiven, some account to reconcile. It's always hindsight that allows us to see the moment we bought the ticket, the pop of the stage lights coming back on, our ascent up the stairs to take our part in the encore.

The evening in the village unfolded without having a proper chance for conversation, so he insisted that we hang out afterwards. Together we hopped off the tour bus in centro, grinning like foolish children. His overt romanticism was innocuous, and my tolerance was bolstered by the drug of impulsivity. Forgiveness is complicated when you're attracted like a moth to the light of beautiful delusions. The warm night air and memories of our high-spirited fun set the tone as we ran to the zocalo together. We decided to have a drink at his apartment, and piled into a taxi.

Sitting across the table from me in his modest bachelor's kitchen, he refreshed my vaso veladora of mezcal. He confessed elaborately that he *had* to stop talking to me because he fell in love with me. I told him how ridiculous that was, that love doesn't work that way.

"Why didn't you fight for me?" he asked.

"Your message said that you never wanted to talk to me again, how could I ask you to change your mind and still respect your wishes?"

We argued casually, as you do in retrospect, the words seeking understanding and not consequence. Our problems were cultural among other things, and as the impasse of our perspectives became clearer, I eventually asked that he take me back to my hotel. I had been playfully dodging his kisses, but I didn't want to lead him on further if he was being remotely honest about being in love. Being lovers was one thing... being *in love* was something I couldn't reciprocate.

As I was getting out of the car he caught my arm and pulled me back. "I don't know why I didn't marry someone like you," he said softly, his dark eyes glistening. I smiled sadly, trying to receive his over-the-top remark but unable to internalize it.

"You are better off without someone like me," I said unequivocally. Deep within me, lament for our fate and for our mutual loneliness was a dull knife, twisting. I really did care for him. But I couldn't imagine how utterly unhappy he would be with a partner like me: traveling the world alone, being polyamorous, autonomous, making my own money... in this culture, these things were simply unheard of. He would lose his mind from the inability to contain me. Moreover, saying something about marriage in this moment was an enormous red flag, a love-bombing hook that could never catch me again. I have endured enough control and manipulation for several lifetimes.

I kissed his cheek a final goodbye and walked through the courtyard to my room. It had been good to see him even if my heart could not be convinced of his explanation, nor his renewed romantic pitch for my affection.

*   *   *

The next day my friend messaged to tell me that he had been her tour guide that day, and that he was hitting on her for the duration. I received her message between ones from him that were sticky sweet romantic, laced with his daydreams of us traveling together. I asked him if he'd run into her. He responded casually, then defensively, then dramatically cut the conversation off to tell me he was too tired to talk. Ugh. Obviously this was a game, and I had no interest in playing it.

He had bombarded me with an outpouring of messages for days since seeing him again, but after this odd exchange, he avoided me for almost a week. Finally, I asked him what was up, and he exploded that we can't be friends again, that I am an awful person, and that he and all his friends laugh at me.

I sat reading his confusing and insulting messages and felt the sting of humiliation. I had to dig deep to not engage in the exchange. I resisted the impulse to defend myself against his projections. I had no need to disparage him to end our friendship, and I told him as much. I wished him love and peace in his heart. And I meant it. It hurt that he laughed about me. Why take it there? So hot and cold, surely all his romantic posturing was as vacant as it had seemed then. But I wasn't prepared to hear that it was a joke, that I was a joke to him. He was only trying to hurt me by saying that. So I wished him even more love... so that someday he would not feel the need to be inhumane to other people when they were not right for him as lovers, as partners, as friends.

* * *

Now I sit here this morning in a coffee shop in San Cristobal comforting myself from the bite of his words, spoken to wound me and make me feel separate. Alone again in this new place, I let them in for a moment. The damage is more than topical, it is archetypical; it has less to do with the particulars and more to do with the patterns. Even as my life energy for superficial social dramas wanes, I still don't always have the resolve not to actively participate in them. My frequency is shifting away from seeking or needing male validation, but applying even minor external pressure distorts my worldview back to the more familiar, damaged one.

The cruelty is also familiar, the tactics of a threatened man. That which cannot be controlled must be neutralized, dismissed, or destroyed. Those who see through your bullshit must be punished. I've triggered something unhealed within him as well, and this is how pain perpetuates. His reaction is because he is suffering and can't connect with it to resolve it. Somehow my heart breaks with his, *for* his, for the collective wound in the masculine. Healthy people have no need to inflict harm on each other. I try to keep my empathic reaction separate from my damaged ego, and I work on them both.

I am surprised at my injury, since I try so deliberately now to move through life without causing harm. I haven't endured an ounce of drama since The Four Month Break Up. That doesn't mean anyone else understands me or is in the same place in their journey. I stay present with my discomfort as I study the rubber band ball of my heart.

I look at the storyline now, wondering what a better choice could have been. Maybe, truly, this falling out was not my fault. Or maybe it just needed to happen. There are lessons in all these things, I just want to make sure they are what I *think* they are. I wish that I hadn't seen him again, that my memories of the previous year were yet preserved in their sweetness. But the reunion gave me clarity, as bitter as chewing an aspirin. I'm still sensitive to getting caught up in some man's projection of me. I'm still one swiftly delivered emotional punch away from my PTSD. What he said was unjustified, but it takes me awhile to sort through and make sure. I am not a bad person, I am not a liar, I don't carelessly sleep around with people, I am not a joke, and I don't deserve for those words to stay within me. I let them hurt, I acknowledge that hurt, and I let them back out. I want to return to love as quickly as I can, but I know it will take a little time. Maybe an hour, maybe a day, maybe more- but somehow soon I will be reminded of my worth, and my loving nature, and my belonging. Those things are not on offer for someone else to take away.

Returning to the same place again… but as a different person. Life is truly an accumulation of moments, each different than the last. This offers the brilliant possibility that our growth potential is not limited by the environment in which we begin, we are not fixed, we are not condemned to our past, we can change completely. But also encoded in this realization is the forward motion of life: there is no "going back" to what has passed. Each moment is fleeting, each person is evolving.

I leave here with a few thoughts:
-People didn't heal me last year, I healed because I was ready to heal, because I was ready to see what needed to be seen to move forward. It awaited me here or anywhere, but I was open to it here because I surrendered. Southern Mexico offers a certain brand of magic, but so do all places if you are truly ready to see. Traveling makes it easier to surrender, but surrender comes from within.

-Some friendships are not meant to last, and that is ok. Learn to let go when it is time. Learn to find closure within if things feel unfinished. You owe people your honesty and integrity, but that is all. You do not have to explain yourself until someone understands you. You do not have to explain yourself at all.

-Some people quietly love and support you whether you are able to talk often or in a shared language. Appreciate them.

-Last year I left this place feeling refilled and ecstatic. This year I leave this place feeling complicated and lonely. Both ways of feeling are gifts, but clearly one takes more effort to unwrap. I vow to sit with all my feelings that arise and give them nourishment and validity. I vow to honor exactly where I am emotionally without expectation of where I should be.

-I had a lesson on feeling harshly judged. I vow to let it have only enough power over me to make me thoughtful of my behavior and choices, including who I let close to me in my life, and whether I value their opinions.

-My resilience has improved but sometimes at the cost of my happiness. It takes me a lot of energy to transmute something negative back into love. I vow to take care of myself when I am doing this work, because this work is important. I will take my time, so I do not compromise my health.

-I don't have perfect integrity, even though it is something I cherish immensely. I still have a hard time telling people things that will disappoint them. Sometimes it takes me a long time to process how I feel, and life is moving along more quickly than that. It doesn't make me a deceptive person, but I need to be careful to give myself more time to think before I respond if I need it.

-I am disgusted by the amount of time I spend on social media and vow to be more present with what is around me. I use this as a crutch against loneliness, but it never helps. I feel overexposed.

-I have anxiety about my upcoming return to Baja. I don't want to lose the beautiful memories I have of that place by renewing them with something less fulfilling. I need to take the time to work on my expectations, my presence, and my boundaries, before that trip. How can I best avoid disappointment? I need to apply what I have learned from coming here again.

-Cultural differences are complex and can't easily be reduced to an explanation. These things are in our epigenetics, in our collective memories, in the ways we were raised. We can try to understand the differences, but we will never truly know how it feels to be *of* a different culture from a personal viewpoint. We cannot escape our own lens even with deliberate objectivity.

-Even as I learn to create an examined and deliberate life, it is a practice. I have spent decades relying on externalities to generate feelings, and it is a hard habit to break. The knots come loose but only to reveal more knots.

Mexico City

I knew it. I knew that the universe would restore my faith in people, and it happened almost immediately. What a gift to end my trip this way. I'm exhausted from lack of sleep and the cumulative excesses in my travels, but I am emotionally resuscitated. I spent the evening in incredible company, and it was salve on my disillusionment from the bizarre and childish turn of events with my friend in Oaxaca. I know that this high will fade, that its just one more dose of medicine I've stumbled upon rather than created. But it's a drink of cool water after weeks alone in the desert, and I let it pool in the cracked clay of my mouth. I don't cling to it, I just allow it to replenish me.

The first time I met this man, we were on a train next to each other from Machu Picchu to Cusco. In my limited Spanish and his limited English, we spent the three hour ride comfortably entertained, sharing conversation and silence with the depth and ease of old friends. We had known each other in another life, maybe. With some people, you skip the pleasantries and just jump straight into the cosmic.

I had always hoped to see him again. In my life and in my travels I have met a precious few who have also plumbed the depths of their souls, wringing the wisdom out of their trials. It is rare and beautiful to intersect with someone riding the same frequencies, and I would go to enormous lengths to indulge an experience with any of them, anytime I can.

Yesterday I rented a room in the city that was more like a luxury flat. After staying for weeks in dingy little hostels with no privacy, I took the afternoon to dance around the entire apartment. I soaked in the clawfoot tub and reveled in the opulence of a bed that rolled on tracks onto an epic, furnished balcony above Roma Norte.

As Mexico City is my friend's home town, I extended an invitation for him to join me. He arrived in the early evening, and our time together unfolded as such a surprising delight: how he radiated happiness and depth and curiosity, how he shared of himself, asked thoughtful things, had such presence. He occupied his space. He embodied so many things that I aspire to be. His company was honey on the wound of severed connection and complicated people.

We talked for hours on the terrace, discussing relationships at length. Love. Timing. Individuality. Freedom. The meaning of life. Other dimensions. How time works. We laughed and blew smoke rings from cigars, we polished off a bottle of champagne and a few pours of mezcal. Inevitably we reached a pause, suspended in the palpable electricity between us, enthralling and delicious. But the chemistry didn't have the usual urgency or insistence. It was weightless, somehow sophisticated.

When the humming of the tension overpowered our philosophical ramblings, we said a heartfelt goodbye instead. It was the best night that I've ever been... rejected? Respected? It's unclear to me exactly how we found the reticence to refrain, but I was left overflowing with gratitude for connecting with someone that valued integrity and friendship over physical gratification. There is a long game in store, perhaps we will be friends for time to come. The elation from making a different kind of loving choice was so emotionally fulfilling that it palliated the attraction. Barely, I will admit. We laughed about narrowly escaping becoming lovers, and it was genuine and good.

Perhaps no one I have ever met has so gracefully displayed this kind of resolve. It wasn't his default expectation to be anything more than platonic. He had a partner, they were monogamous, and he honored that. I respected that so deeply that it triggered a bit of shame at my own assumptions.

It dawns on me that I have been conditioned by the types of people that I've taken as lovers, the types of men that I have had close to me, the people who have hurt me. In a way, *I* was the one that had an expectation for the evening... I must have believed, at least on some level, that I would potentially sleep with him. Otherwise, I would never have invited him up- I'd have chosen to meet anywhere else, someplace that I could escape. I've spent a lifetime protecting myself, sometimes poorly, from predatory and opportunistic men, and apparently that has molded my behavior. Letting someone close to me, even in a simple social setting, is preceded by some internal calculation- one where my intuition and ability to control the environment are factored against my naïvety, desire to relate, interest in the person, and willingness to engage in an experience. It is a risk analysis, and I don't always have enough information beforehand. Am I subconsciously pre-deciding whether I could be physical with someone, so as not to have an encounter where my choice is taken from me?

Last night made me unpack my own presumptions, and I see them as unenlightened, crudely resigned to how the night could go. Before he even arrived I had already committed the act in the deepest reaches of my mind, just in case it happened. This is not a self-depreciating behavior, not a casual attitude about intimacy, not a judgment against men, and not an evaluation of an individual's tendencies- this is a response to trauma. And it is rising into consciousness, asking to be healed.

If I want to conduct my life with high authenticity, I need to adjust my energy to reflect what I expect of people and what I *want* to happen- not what I assume, or fear, or that to which I might passively submit. I need to stand taller in my ability to manifest my experiences, and reclaim my agency as both a woman and as a person. And I need to learn to trust myself again by leaning into my intuition about people. I want for myself the deeply embodied presence that I saw exemplified before me last night. I want to carve out a life of intention, deliberation, discernment, and grace. Once again, I find myself prying apart the ways I have been programmed to decide whether they are serving me in creation of the optimal life I imagine possible.

Such a potent reflection, delivered in a gentle way from an admirable person. I have so much to learn from this individual, and I think he does from me as well... there is such an easy mutual fascination. This one, I want to keep. I want this human in my life. I want to be surrounded by people who inspire me.

As the night closed, we hugged the hug of a thousand arms, wrapping heart to heart. His footsteps faded down the hardwood staircase, and I went to the balcony alone but no longer lonely. My last night in Mexico. My last moment in this version of myself before the expansion of my heart from this love, this lesson, this new surrender. I turned the speakers up and like magic, the playlist was in perfect resonance with the essence of it all.

I danced beneath the white-silver waning gibbous moon, among the buildings rising up like gigantic trees from the urban jungle, among the infinite perspectives that played out alongside mine in the murmuring metropolis. I danced to heal the little wounds of contrast and mortality and timing, the bittersweet dance of goodbyes large and small. I danced to thank the universe for reminding me that more love is always the answer. And I was not alone at all. My liquid gray shadow on the stucco danced with me. The city pulsed in time with the blood in my veins. Everything was alive and in flux here within and around me, and I was grateful, sad, complete.

Home

For me, unity with another person has often come with giving up a piece of myself, yet being in love and sharing a vision with another has made me so willing to pay this admission fee. As I cultivated a stronger sense of self, I adjusted what I was willing to give away, but partnering still requires a degree of self-sacrifice and compromise by its nature. There is a balance to be challenged into evenness. With greater integrity comes greater boundaries, and with solid boundaries what is offered can be given more completely. The adjustment means that what is exchanged is deliberate. It's a refinement.

I crave the kind of connection where people dive in head first not knowing where the bottom may be, but I also require the kind of autonomy that nourishes my place in the world as an individual. For a long time, these things seemed mutually exclusive to me, a choice between integrated partnership and self-development. I can see now that these are not opposite ends of a spectrum. Dissolving yourself into unity may seem like a romantic sacrifice, but codependence undermines one's core vitality. How is this such an overlooked failing of our culture? It's the basis of every Disney fairytale we are sold since childhood. And it's the default mode of operation modeled by many women, and expected by the men. Nearly every woman I have ever known has, to a great degree, fit her life around that of her partner, wedged it into the gaps in the periphery. A life like weeds growing in sidewalk cracks. We have to teach ourselves not to do this if we're born into it, retrain others how to treat us. It is so uncomfortable taking up space when you aren't used to it, but it is critical to the heart of a being, learning to cast a shadow without apology.

Optimal relationships are created by partners who encourage personal growth to the highest degree while fostering a grounded, loving connection. The balance is a moving target. It requires deft communication. It requires the deciphering of an emotional evolution as someone changes and recalibrates their needs, either in indiscernible quantum adjustments or giant leaps. This kind of partnering takes bravery.

The only person I've come close to sharing this with waits for me, or doesn't, at home. A home we built wrong and then ruined. A home that I dream about fixing as I run away from it. There is hope, I think. Every ounce of healing that I find out in the world helps me come back, bit by bit, believing in rebuilding. Every new layer of Self that is assimilated is a step toward wholeness, a Self that can actually *be in relationship*.

Nin's words echo: *We do not grow absolutely, chronologically. We grow sometimes in one dimension, and not in another; unevenly. We grow partially. We are relative. We are mature in one realm, childish in another. The past, present, and future mingle and pull us backward, forward, or fix us in the present. We are made up of layers, cells, constellations.*

I am grateful for the relationships I have had before, for those whose lives have collided with mine. Because of them, I am more capable of love, more in touch with what I want and don't want, more discerning in my life. For anything they took from me: time, heartache, frustration… the gifts of resilience, perspective, and introspection have enhanced my life to a greater degree. *What they've given freely and what they've left behind have been formative beyond measure.* We are all emotionally where we are right now as a result of the patterns of thought we developed in relation to our perceived place in our environment. How we feel about the people we have loved contributes to developing our relational identity, a lens through which we now view our ability to love and be loved in this moment. For all the mistakes I have made, they have led me here. Closer to truth, closer to home. May I forgive and be forgiven.

What new beauty will today breathe?
What new loves will fill my soul like fresh air?
I am more alive now
Since I met you

What hopes could I have to see you again
To jump into the waves you make
On the shores of your life,
To play as children in love with the world

It's enough to know
that you exist
To make me believe I can swim
a little farther out
where I can't touch the bottom.

My heart aches
Whispering the first name she knew
Rapping on the door with soft, deep persistence
Love,
I am here to meet you where you live.
No need to permit me,
I am already within.

What choice does she give?
I am destined to surrender
She sends me You to request this,
you to inspire me to unbolt my own resistance.

I send the letters within as
I open like a seed in the warmth
of sweet messages from the sun.
All my gratitude to you
For helping me re-member,
For witnessing with grace
The design bigger than myself
That moves me completely.

What strange spark have I been gifted?
For a friend who needs not to be kept
to be adored
And my own love that blossoms in the light
of my attention.

Every day offers something new to inspire growth- as a human, as love, as love learning to be human, as a human learning to be love. I fight with my small self, the one who feels ashamed of how many things I did not know before I learned them. Every single day I feel the gnawing ache of growing. The small self unfurls the scroll of my mistakes and spurns me with criticism.

But there is a deep quiet building within, some ever-curious part of me, the one beyond ego. She looks at the scroll inquisitively, compassionately interested in the complexities of aliveness. It is she who comforts my child-self when I fail, it is she who leverages the mistakes with gentleness into a new way. She reminds me that I am worthy of my own forgiveness when I am naïve, when I misread others, when I act out of fear, when I cause pain. She reminds me that we come here deliberately to this exact life- we borrow these bodies to learn. And we need not be perfect to be loved and loving... in fact, love is always messy, unpolished, and evolving. Love is always learning.

I offer myself non-judgment and forgiveness in reverent humility, and I offer my imperfect self as I am to those I encounter in this life. May we meet ourselves and each other with empathy and gentleness as we fumble along in our respective journeys.

There is no cure for me because I am not ill. I should be careful not to ascribe all my idiosyncrasies to pathologies. I think I need a break from cognitive behavioral therapy, as it is distorting my self-perception as I heal.

That which triggers us is a doorway to empathy. If you find yourself in judgment, pause there. What about this person is a mirror for something unhealed in yourself? By offering acknowledgement and naming the source of discomfort within, you can deepen your acceptance for yourself and others. This kind of practice can open us into compassion for our archetypical shortcomings as humans.

If everyone was already a perfect being, our social constructs would be devoid of contrast and meaning, and we would be deprived of distinction in our relationships. If we were all fully formed, the we would be denied the divine state of growth-the very thing at the cusp of what makes us human. Applying our consciousness to our inner worlds to evolve emotionally, adapt deliberately, and make choices in our best interests (sometimes defying our biological imperative), are potentials allowed to us because of our sentience. The feeling self is our means of transcendence.

Increasing our introspection requires creative practices to open the aperture between the self and the inner witness. This is experiential and cannot simply be intellectualized. Investigations into one's psyche can be contemplative, peripheral, ethereal, religious, or induced by trance, trauma, sex, meditation, illness, or substances. I believe the empathetic response is our natural and highest state, revealing itself freely once the survival-based, ego-driven animal mind is peeled back or taught to relax.

We uncover our hidden true nature, sometimes, with a sort of emotional sonar. In our lives we encounter challenges of varying frequencies. If we redirect the waves within to see what they bounce off of, the reward is a clear image- the edges of the real self, expanding in its course of evolution, waiting to be honored and actualized with our attention.

Retreat to the sanctuary of inner wildness,
That quiet highest altar
that waits behind the curtain of language and
narrative of relation-ship
for this indulgent reunion of soul and self.

In this place
Holy and unpolished
I am restored to my universal state,
The birthright of divine connectedness
That defies the grasp of analysis or obsession
And sublimes straight into knowing.

I'll surely succumb to an early demise
Burned alive in passion
Either from the one within
or sparked from a great love.
My heart is a heap of gasoline-soaked rags
And I chain smoke on the stoop,
waiting for you,
waiting for me.

What but longing
Could keep us so ecstatically upheaved
Untethered by all but the hope
to realize something deliciously improbable?

Bring me the horizon of myself
So that I can challenge it with
all that I have yet to become.
There are worlds within
To which I hold but one of the keys.

I can be so tragically codependent, so knowing this, I attempt constant conscious growth into the expansive open love that is the true nature of the heart. I don't pretend to be perfect at this- not even close. Sometimes my life is clearly the battleground where what I *feel* and what I *believe* come to wage war. But every maladjusted coping mechanism I repair is a weapon that I learn to drop, ushering long-overdue peace into my interactions with others as I find it within. My armor piles at my feet, and I meet the world with the refined resilience of softness.

Non-attachment is difficult when you are hopelessly romantic... but every time I am able to let go and give up ideas of what love needs to look like, or what love needs to provide for me, I am met with deeper and more complex, more beautiful versions of human connection than ever before possible. I am cutting the cords with toxic relationships, repairing those that are worthy of investment, and attracting better friends as I become a better friend. As I heal the fear-driven default of my animal, I am finding the limits of that mindset shattering apart, one interaction at a time.

True love is freedom. True love is allowing things to be exactly and only what they are. In deeply respecting someone's autonomy, you offer them the ability to value you intrinsically (and not because of a set of expectations). This allows the relationship to assume its own natural shape, taking up as much space as it grows into. It blows the ceiling off of vibrational constraints that come with labels or the commonly imposed timelines or narratives. It permits those coming together to show up as they are, standing on the foundation of showing up first for themselves. In that, a rejection by another is simply an acknowledgement of a mismatch in frequency- not at all personal.

So many of my disasters of the heart have been catalyzed by a fear of abandonment by others. For so long, I did not understand that my self-abandonment perpetuated the very thing I was most afraid of. Every micro-rejection felt like an erosion of my worth, which made me cling tighter to externalities all the more. I am finally seeing this clearly and working to change it.

Any time we become afraid to lose someone, a way we feel, a routine we have come to expect, or our construct of another, we subconsciously close our hearts and clench our fists in the false hope that control can grant us some type of permanence. This response is protective in essence, but stems from fixed-mindset lack-driven thinking. It doesn't rest in the reality that everything changes. It doesn't honor our high self, it doesn't value others as being on their own evolutionary journey, and it doesn't help us build quality, grounded bonds with our romantic partners, friends, or family.

Only by offering our own bare honesty and surrendering into trust (especially when it is hard) can we know what it feels like to cultivate expansive, emergent, unbridled communion. Only by allowing people to flow freely in and out of your life can you value them in the moment-by-moment reality of relationship, and let go of the mental constructs that freeze others in place like objects. Nothing is owed, all is freely given. Discernment and self-worth fortifies our own boundaries, and active respect illuminates theirs. You have to love yourself first to have anything to give. I see that like I never have before.

To know someone is to get to know them over and over again. To love someone is to do this deliberately, and because you want to.

Looking back at my pictures through the years tells an uncomfortable story... that even at my happiest, those moments that were chosen to capture and savor, I could not access sincerity. My smiles and poses more often than not are exaggerated, melodramatic, falsely saccharine. It makes me sad to realize that after all this time, I have been using cynicism and silliness to escape the ways I don't feel worthy of real sweetness or gentleness. I want to make a promise to my heart that I will, from now on, hold the space necessary to allow true, genuine, happiness should it find me.

It's amazing, my emotional reaction to just writing that.

I want to be well-adjusted and authentic more than anything I've ever wanted. I don't want to wall myself off from sincerity with posturing or ridiculousness. I want to allow the honey of love and abundance to flow into my life. I am no longer ashamed to admit how much I hope for something good.

I escaped into otherness
Liberated from a Self
and yet communing with all the
Selves at once,
Ones I have been and ones I know
only from Somewhere Else,
From beyond memory,
From the rich universal amniotic.

Free of limb and burden,
Serpentine
Flowing with the cool river,
As the river,
Pouring with ease and relief
Downstream,
As was my design and my destiny,
My body resembled its path
Winding over terrains that all felt like home as
I passed through them.
I was moving, moving,
and always I was
Exactly where I was supposed to be,
Effortlessly.

Then...

Free of bones and weight
Free now of gravity or confinement
Twisting
Rising as ribbons of ritual smoke
Unfurling up in tendrils
Lighter than air
Lighter than souls
Lightened of the heaviness of
Belonging only together as a being,
Together under the relationships
Of all that makes a Self
Released of impermanence and structure
into the expanse of an Everything.

Now surrendered of form completely
Swirling in the singing bowl
Tumbling between fragments of oxygen
Reverberating
As the floating trill of bells in the wind
As the essence of freedoms
Escaping into a finite instant,
A song being played,
Exactly and only
What I always have been.

I Unkeep the body I live in,
The attachments to memories and narratives,
I unkeep my manifestations of selfhood,
of importance carved separate from the Wholeness...
There is no longer a need to explain myself
Or a desire to show anyone all of my faces,
The forms I have taken,
The ways I have dissolved and recrystallized.
I am and was always part of the Everything...
The distinction is only whether we
can see one another now
as our true selves.

I Unlearn ideas of legacies,
Of differentness or specialness,
The things that come from the false wound of Unbelonging
And that keep one too weighted down to experience
the buoyant elemental reality of a soul.
We are only moments in time,
Lit matches meant to burn.
I can feel myself on fire now,
Can understand what it means to be alive
Billions of candles alight together
like a constellation of brilliantly combusting stars
Buttressed in togetherness
against the backdrop of
A vast and unknown darkness.

I'm beginning to understand that I am the being that lives within this animal, but I am not entirely or *just* this animal. I am at once exactly her, but also somehow beyond all the limitations of circumstance and experiences that have shaped this body and mind. There is a me not confined to the invisible edge of her sensory boundaries, I am held here within her by a single silver thread. I love this girl that my soul wears like a winter coat, because through her I understand what it feels like to be warm against the outside cold, what it feels like to affect things with my hands and voice and heart. I love this girl animal that loves other people so fiercely, because her journeys in connectedness have given me the chance to experience the range of human emotions. I am her and she is me, and yet, she is individual and I am yet uncleaved from the whole. We are learning each other, and maybe, just maybe, that is the whole point.

Day and night is not a thing that happens. It is a place.

A circle, a cone of light abutted against the everything of darkness, a line in space, a threshold we move back and forth across as we spin. The sun drags us along in her wake, and the most important reference for us in our flicker of existence is the regularity with which we visit the shadow of ourselves.

"But I just want to *keep*," she said softly, sadly. It would seem she was talking to herself, but I knew she was talking to me-the one who lived within the shell of her, the limits of her. Her voice cracked with knowing, and the levies of her eyelids trembled with their burden.

I'd seen her in the mirror nearly every day of my life, of our lives together, but only recently had we begun exploring our relationship. Only recently did she say these kinds of things to me or ask for my council. We tiptoed on the spidersilk strung between us: a timid understanding, a fragile trust. This was the kind of love that was hard to let in after all the wariness born from allowing bad loves to poison her. And we knew. We knew it would all somehow be okay now that we had each other. Still the recognition of each other meant an awareness shift to a higher perspective, and with this, the embodied realizations of her smallness, her mortality. There was innocence in her response, a childlike sweetness in her surrender of the delusion. Tears breached their threshold and spilled in quiet, cleansing rivulets, unencumbered by the hiccups or weeping of resistance.

"I know, my dear animal," I said to myself, to my body, to her. I hugged her from the inside and patted her thigh to comfort us. I knew exactly what she meant, she wanted to keep something, anything... this body, this moment, a love, a story, this life. Any of it.

"But we can't keep, we aren't meant to keep. That is what we came here to learn."

I have been struggling a bit with whether I want to put my work *out there*. When I reread some of my stories I feel my ego catching on all the parts that are not smooth yet, on all the mistakes I still make, the ways I am still addicted to the world, addicted to the pain of my own endless craving. But it's *real*. So I leave the entries raw, without paring them down or addending them for how much I learned afterwards. They are strange works of art in the gallery of my soul- a series on Becoming, that I walk through with aching overexposure and compassionate objectivity, pondering what it all means.

There are so many ways this journal is only pieces of my heart, scraps of stories. Part of me wants to go back and finish the tales, sand them and varnish them with hands now steadied by the lessons. My self-image aches in the commemoration of naïveity and immaturity, of graceless fumbling through a life. I want Soul to have written the whole thing, I want to be wise. I want to add the way that I felt the next day, or how the chips fell after I poured my feelings out to the pages. But life moves forward too quickly to fall that far behind, and I always feel differently about something once I write it. I remind myself that it is not my goal to have a complete historical record, but to write when I feel moved. There is honesty in leaving it, but it's so acutely vulnerable.

Sometimes my clarity feels perfect and I feel the light of universal truths emanating from within me, a dismantling of my illusions- and then the very next moment I am thrown back into another lesson for Animal. It's almost always about romantic love, self love, belonging, and boundaries for her. She gets so stuck on these things. She has come a long way, and I love seeing her grow and listen more to Soul. But the switching back and forth in perspectives is labyrinthine. I reread things I have written and see the two sides of myself taking turns at the helm of my life- curating my experiences, authoring my stories and poems, making my decisions, forming my thoughts. I am so clearly both this girl animal at the mercy of her own emotional hurricane, as well as something else entirely- a spiritual being, wiser than my relational self, capable of existential clarity and transcendent consciousness. I ache to sit in the throne of intention and integrity at the core of my being, unifying these aspects of self into more consistent deliberate living. But I am yet undoing lifetimes of trauma and cultural malconditioning, tossing in the wake of residual karma even as I begin to witness the perpetual stillness that occurs at depth. I am changing, but still suffering the consequences of things already put into play.

I remember compassionately that *I am only here to learn*. Acceptance of this makes it easier not to want erasure of that learning: those things Animal does in passion, desperation, and survival. I am witnessing myself being human, and there is the paradox of divinity contained within that if I can surrender my shame and open to the complicated beauty of my bare truths.

Now that I am thinking about myself like this, as Animal and Soul, I wonder- which one is sick? Which one of me has bipolar disorder? How can I understand my diagnosis in this framework?

I think my mental health is not in some permanent state that has a label, nothing is so black and white. It is not like I have an incurable disease, it is more like I have the flu, or maybe like I am prone to catching the flu frequently. I do not feel like I am always compromised by my mental health issues now, I feel like my mental immune system was destroyed by trauma, and now I am recovering. This is not just euthymia, it is restoration of my healthy natural state. If I go back to a therapist, could they peel the label off? It doesn't work that way, I'm sure. Even though my own self-reporting got me a diagnosis, it's probably not common for therapists to declare someone healthy again. To do so would be to suggest they were wrong, or that bipolar disorder can be acute. Anyway, what does it matter what anyone else thinks when *I know* I am healing? *I will not get stuck here.* I am not the same person that I was eight months ago when I started writing this book, or even the same person I was last week.

I think Animal is the one that gets sick, that it is within her that the issue manifests. When Animal can't hear Soul, when they are separate, Animal succumbs to the consequences of this disconnectedness. Her poor coping and inflamed or overreactive emotional states take over, and this is when I feel episodic. It is the body that suffers, or the body in conjunction with the mind in a negative feedback loop. When I am too entangled in the world, I am reacting, dysregulated.

When Soul is strong and is guiding Animal, I am in harmony with myself, and I feel grounded and fine. I know when to rest, when to eat, I know how I feel, and my feelings are proportionate to reality.

I feel like the way to ensure my mental health is to strengthen Soul so that Animal has a guide to purpose, centeredness, and connection. She has reasons. She is balanced.

Soul *could* be sick, and I give this some thought. But Soul is pure, elemental, my High Self. I think she is untouchable, *La Que Sabe*. Animal can't injure her, it is only the other way. It is only when Animal can't or won't hear her that I suffer. So I need to work on Animal's listening and Soul's voice. The only way to health is by cultivating an undivided life, one where my intuition is honed and my actions match my guidance. The sensitivity that has made me prone to mental injury could be the ally that helps me feel my way into wellness.

When I am small,
Smaller than my Self,
I sit with the stars to
be reminded of
How this dance works.

How many
Are on paths fated to intersect with mine,
Unseen gravities
pulling us into collision,
to that instant we are born into the world of each other?
Let me stay the course, knowing
Everything is on the way.
We are never alone for long.
Let me be humble.

I can not presume or predict
which realities will be deflected
or absorbed into the ink
and which will circle around mine
with perfect balance,
caught in the beautiful grip of Otherness.

The end of the year as we understand it is really so arbitrary, but nonetheless I feel a contemplative closure as the squares on the pages run out. I do become more pensive this time of year. As the days shorten, my mind turns to wrapping up the stories I have lived into a collection, making my "Best Of" mental slideshows of memories and moments and pictures. I am quickly bored with the unrealistic slant of only remembering the highlights, so I attempt to compile the beauty of the mundane as well: all the things that didn't get recorded in photos. I wonder how many times I cooked meals for my friends this year, how many times I danced, and how many times I have stared at the stars with renewed awe. I wonder how many times I said *I love you*.

I've lived through 36 trips around the sun now, but we are culturally conditioned to see them as linear blocks of time, masculinely carved into neatly-tiled equal segments of life. I imagine the years as timelines stacked horizontally, calendars with a photo for every day. Maybe it's more like a movie reel still playing, spiraling behind us unfurling into the past as we press on.

Some years seem to slip by under the radar, the ones where goals were not yet met or nothing major changed: the third year of college, the year before I got a car, the in-between years of status quo. There is a block of childhood that I think of fondly. I remember playing outside every day, but have few distinct memories. It's weird how we remember the past as a vague amalgam if we don't record it externally to ourselves or have strong emotions embedding it more succinctly into our minds. If I think a bit harder or dig for photos, I could come up with a title for most of my years and surely find (or create) more importance.

Years leave all kinds of different impressions, some make us different than we were before, changed ever after. They are etched into us like black tree rings from the wildfires we somehow survived. There are the years marred by loss: the year Grandpa died, the year I got sick, the year my parents divorced when I was a kid. There are the years that stand out as kinetic and potent: the year I split my life between two men and two cities. The year we moved across the country. The year I began traveling.

The years that hold the weight of negative space are more complicated: the last time I smelled Grandma's Thanksgiving dinner cooking, the time before I understood my depression, the time lost to alcohol, the last time we kissed like we meant it. These things don't really have marks on the timeline until afterward. We are really always annotating the margins on the catalog of our personal flipbooks, if not reworking them entirely.

I color the sad boxes flat stratocumulus grays and the brave ones a fiery vermillion. Textures and hues for every emotion. I take a step back to look at the the bigger picture encoded in the collage of pixels, as though my memories could be condensed to one of those photos made up of hundreds of other photos. Maybe the days and moments are fractally expressed- maybe the detail is infinite, revealing itself at the scale one chooses to view from. How the symphony of shades changes across time, the smudges of complex feelings stretched out across days and months; the sharp contrasts of traumas and joys spiking boldly- bright, emotive paint flung onto the abstract of my life.

My sense of time feels bent. None of these things that I have lived through seem like they fit into an instant, a day, or even a highlighted period of time. The colors never become separate or homogenized, nothing is discrete. So many of the stories don't even feel right to condense to words. They meld and fold together, they overlap, they play on top of each other like instruments in a song. They end and yet remain. They become part of me. And since I am still here, I *am* these stories. My memories are cellular. I am still living all of the things that I ever did and all that happened to me, time and timelines and tales all brilliantly woven and tangled into the tapestry of my Self.

There is no combination of poetry or art or photos or songs- there is nothing that can come close, especially not a calendar- to recording the net sum of what the world has given and taken, how I have *loved*, who I have become, what I have dreamt, what has shaped me.

Time is fluid; the Self, dynamic. The layers of who we are can never really be reduced to snapshots in time, and time can not be confined into convenient chunks to be filed away or lost.

Still, I want to play the game. We are all on this ride in space, and we are approaching the line again, the one we made for reference... what has made the journey around the sun meaningful this time? What gives us hope in our next revolution, makes us circle the boxes that stand as proxy for points in time, places in the cycle that we expect to experience? What colors do we imagine for the painting of our unpromised moments? This time of year seems to give people a sense of scale that is easier to access only in endings.

I wonder how many times I took out the trash or grappled with existential confoundment. I wonder how many miles I hiked and how many tears I cried.

It's been a good year.

Dreams:

1) *Nuclear war. I was at Kelly's house (but it was like a trailer, maybe near Grandma's house in Fruitland). I was holding Loki. I ran outside to the barn, but the ground was liquifying. I hunkered down to protect Loki, waiting for the shock wave. The wave never hit, but we could see the mushroom cloud in the distance. I remember thinking that since we didn't feel the shock wave, we were in the zone that would be likely to survive the radiation.*

2) *I gave birth to a baby with one eye missing. Easy birth. I had the baby on a towel on the couch. I pushed once and he came out, tumbled into the floor. I felt the birth entirely, the movement of a body coming through my body. People were watching, but it felt like family, not doctors. The baby was bloody and missing his eye- the eye socket was a just a hole, a birth defect. I didn't remember whose baby it was, counted backwards through the months, thought it could be P's. I was sitting at a table talking to a lady when I birthed my placenta onto the floor. I remember that I no longer looked like I had been pregnant, I looked normal.*

3) *I randomly saw P when I was traveling. We were in a pool. Maybe in Thailand?*

4) *I went to a place- maybe a park or courtyard- with two large concrete "cups" or broken eggs with water and trees in them. They were in ruins. I was lucid, and I knew I had come there to take something back to the waking world for my altar. There were large purple snails in the water. I was running out of time- Andy and I had to catch the bus or otherwise bike an hour back, and I had just given birth so I didn't want to miss the bus. I tried to grab a snail but it dissolved. I grabbed a striated rock and put it in my pack. The dream shifted and I was instantly back at the meeting place- a house or destination that had been pre-determined... The baby turned into a brown dog named Josh that we loved and had long kept as a family pet.*

5) I was at a bar with Hila. A man was hitting on us, and we tried to play dumb to get him to go away. I had just given birth, and wondered when my milk would come in.

6) The baby turned into a calico kitten with both eyes, named Josh.

How do we ever really know we are seeing the real person, the one behind the mask?

My heart is in a vice, as a rapturous memory is threatened with the dark smudge of disappointment. I try so hard to take these mental souvenirs with me, press them like plucked leaves into a heavy book, preserve them through the lens of the story I lived during them. But retrospection bestows a clawing chagrin when we have changed how we perceive someone. I feel the cold walls of a cage of my own making- a childlike willingness to find magic in ordinary men, only to be confounded later by their abrupt reveal of insipidity. I am gripped by an impassioned nostalgia and churning desire to seed a world steeped in myth and mystery, but the red blooms flagging in the breeze are the stuff of unshared dreams. The false gods crumble and take to the wind as ash. I am so often alone in this garden.

So where do I put all the unforeseen expectations I had of a friendship when it now seems that I am standing with arms outstretched toward an imploded construct? Did I imagine the connection?

It doesn't ultimately matter if he was deceptive or dishonest with himself when we shared excited conversations of keeping in touch always. The disconnect I feel now only arose as we haven't spoken since, an unforeseeable twist after being spellbound in the constructive interference of an aligned frequency. My ache oscillates, seeking a place to land: is it worse that I lost an imagined potential with a friend, or that I face the cutting recognition of my willingness to see what I want to see in people?

Maybe we never really know the truth, because it isn't singular. Maybe there is no consensual reality, and my story only tangentially rests against his. So much of any relationship is steeped in our assumptions and projections on an effigy of the other that we conjure in the mind's eye. Other than that which my imagination can speculate, I have no way to guess whether I made a mistake, whether I misread the energy, whether he mislead me, or if the lovely things we said expired in the light of day. I still long so painfully to be understood and cherished, and I recognize in the shadow of this experience that I have no close friends at the moment- literally no one who asks how I am.

I think back to our meeting, kicking the dirt for any evidence of misinterpretation. We divulged so many deep truths: vulnerable admissions, innocent hopes, shrouded fears, stories of adventure and misadventure alike. My intuition felt honed, my heart was open, we were both so relaxed. I was very simply grateful for his company, and he for mine... we were in a sort of suspended reality outside of time. I didn't dream it, we discussed it. I didn't need anything from him, didn't need him to be a lover or a daily friend. But after the closeness we shared, I did think I'd hear from him- that he'd casually send a message at some point.

It is so hard to know with people you meet traveling. Some of my best and biggest friendships combusted into immediate greatness from shared moments of awe and escapade, and some people that I thought I'd see again and again vanished from my life as mysteriously as they arrived. Oh, my poor heart loves to attach to the good ones, the ones that feel like they shift the bedrock of your thinking and reinvigorate your way of being, for a moment or an eternity thereafter. Memories of kindreds wash over me, a swirl of faces and places, bizarre circumstances, trials, joys, exuberant aliveness; and this fills me up again. I send them all love wherever they are now. I am so glad for getting these intersections with spirit, these beautiful coincidences of human connection when they come.

I decide I am okay with not knowing what happened, I set it free. The Ho'oponopono holds me: *I am sorry, Please forgive me, Thank you, I love you.* I don't want to poison the memory with longing or shame. It felt wonderful to connect, to let go, to feel seen and safe and exhilarated by possibility. And perhaps, that night, our exchange held all the sincerity that I perceived. I don't need continuance, only the self-assurance that I know what I felt, and that it was real, then.

Reflecting back, I know in my heart of hearts that I didn't manufacture that feeling- it happened. And that's wonderful enough on its own to record as an experience of beauty, a memory that I can still cherish. This clarity grants me peace, an anchor point of validation, and I realize that even as lonely as I am, being able to trust my intuition is far more important than the outcome of a friendship. Rejection is painful, but not likely personal. Delusion is excruciating.

If my mistake is ever that I am too much for someone as I sit in my truths, I can't mourn the falling away of those not there on their paths. If my mistakes are bigger, I hope that I can be forgiven as human, as trying, as learning. Let me whet the precision of my senses and trust them amid the shifting sands of translation.

What do we teach each other by disappearing? We construct lessons in the place the relationship might be. We can inject compassion into the situation from where we stand alone. We can "forget with generosity those who cannot love us" as Neruda said. The marble of memory that we hold in our palm is ours alone to shoot or pocket, curious in its compact completeness, infinite in the joy it can still bring, significant or not as we deem.

Who have I disappeared from, and what story do they tell? I am so sorry to those I have let down or left wondering.

Let me *be love*, and recognize it with all its different faces. Let me *let go*, and let love flow through me. Let me flow through my life with grace and a lack of expectation.

In my mind I imagine the beach
where we will make the mistake into a disaster.
If I meet you there, I have already decided my fate.

I have already decided I love you.

Write the mistake down and
make sure I want to make it.
Crumple it up and
Put it with your version of love in a bottle,
throw it into the sea.
Throw it hard but
my arms are weak and it drifts back
to the shore.

The danger of you, my achilles heel.
The danger of you, dividing my heart
or pulling me into a world that I don't own.

But the red thread has already tangled itself.
This one like a fishing line
And the others stuck like a twisted net,
like a twisted noose.

What good is my freedom
if I like the pain of the hook?

Strings on my fingers so I don't forget
Strings on my soul so I don't get lost
Tired is the heart that thrashes against the lines.

Preemptive consignment to the likely fate
Unfurling with the weight of my life
Like a spool of ribbon that slipped
from the counter
unfettered,
now twisting as it unravels,
surrendered to
its movement down through uneven forevers.

From the cliff I jump, not remembering
the universe that
tethers me or the life that pushed me.
I let go and fall
the fall of those who consent to gravity.
You only believe in it completely
When you test the edge, teeter
Catch me or not,
Here I come.

Cloud of bad luck around you
And I won't go ruining your life
Twice over
With the spell of freedom that
floats like perfume
preceding my arrival

I will be the death of your fantasy
Or
I will be the death of you, love
Let me choose for you the kindest way.

Those I have met and loved:
The gypsies and mermaids
The healers, the hunters, and wanderers.

Kindreds created from shared moments
When life was most potent
With awe and possibility
And we were small there,
With each other.
Witnesses.

The truth is I am tired of
Small loves
Half loves
Sometimes loves.
I thought that I could take a handful of those
And stitch together one big love,
The love of my life,
That never seemed to arrive at once.

But it turns out
what I really needed
Was to love myself.

So now
So cautiously within me,
So barely a blue flame flicks in a nest of
Lint and tinder;
Everything within me tells me
To mother this spark.
And everything that doesn't fuel
This One
Doesn't seem
Like love at all anymore.

The universe bestows abundance on those who learn to see abundance, this I know for certain.

I have met very fortunate people whose hearts are yet marred with envy, loss, and fear. They are unable to internalize or employ their amassed resources, their luck, or their human capital, and they are disconnected from any happiness that could arise from it. Conversely, I have met people who have endured impossible hardships, while embodying reverent gratitude for each renewal of breath they draw. Whether we are blessed or cursed is always a matter of perception. *What is enough?*

I look back at my years as an adult thus far and realize I have spent much of my life empty, waiting to be filled, perhaps *believing* that I perceived abundance, but I was unknowingly blinded by a hunger that betrayed my desperation. I filled the void within myself with the unnourishing fruits of materialism and a throw-away heart doomed to love hard and lose hard. I built and buttressed empires of self-protection, but operated always as though I could never really win. I believed that my lovers would leave me, and they did. I believed that I needed *more* to be worthy- more degrees, more cars, more shoes, more smiles, more lovers, more *something*. And maybe I believed that once I had this "more" that my real life would begin, the one where I was enough. The one where I was accepted and loved, and no longer had to hoard and protect and struggle against the current that rushed on to carry away my hard-earned pile of sticks. The richness of life was Out There- a harvest to be gathered, tethered in relationship, fenced off from thieves. The wolves eddied in the fringe, ever-present, waiting for a misstep. That which I had gained for myself in the waking world was always on the verge of being taken.

I bolstered myself outwardly with the lavishness of things, with the jealousies we are told keep our partners from straying, and a wrench-tight grip on the measures of my success... but inside the walls, I had trapped myself to live with the fear of scarcity. I was operating within an unseen program, that familiar conditioning that teaches us that we are in an endless competition against everyone else for a finite amount of love, approval, resources, and worthiness.

It took me many years to understand the nature of abundance: it is not the harvest you've stashed in the silos, it is not the loves that you train to reassure you... it is the core belief that you can always plant and harvest enough to feed yourself, it is knowing that you have the power to generate love by being loving, it is seeing the world as beautiful despite its cruelties. Clinging tight is a means of survival in famine or flood, but to grow corn you have to give those precious kernels to the dirt.

Nothing generates gratitude quite like hardship, so perhaps the luckiest of all of us are those who have endured and reframed our tragedies. If my own losses have been the price for how rich my life feels now, then they have been the greatest of my blessings.

Abundance is the reward for integrity tempered with humility; it is a model of self-trust, of being at peace with the uncertainty of life, of manifesting positivity by delighting in that which already exists. It is knowing that there is much to be thankful for no matter how dire our present circumstance; that despite how little we can control in the world, the only thing that needs our hands on the reigns is our own heart, so that it is reminded to *be love*.

The more we loosen our grip on the exact shape we expect our gifts to take, or the exact nature of the blessings we are willing to accept, the more we allow the profusion to flow around us and within us, offering us an endless feedback loop of enough-ness that sustains us even in the longest of winters.

I wonder if the machine can even be changed from the inside. It's the most taxing thing I've ever convinced myself that I was attempting, and on the bad days I recognize the futility of the endeavor. You either become the machine, you become as hard as the machine, or you get ground up by it. The only loving dissent is opting out, ceasing to subscribe to the whole paradigm- otherwise you are inevitably poisoned by the program, slated to lash out against your own invisibility, the injustices, the madness of it all. Right relationship with livelihood is crucial to wellbeing, whether or not your life purpose is bound to a job. We cannot give the best of ourselves to something we don't believe in wholeheartedly. If we spend our life dispassionately spilling our time and life energy onto infertile ground, it is an assault to our soul.

There are other ways, I have seen them, and I ache for better work, for a reassessment of priority. I want to do something meaningful, I want to help people. I have a lifestyle that makes this current path hard to walk away from in the meantime. I have so much freedom in some regards, and that has been critical for my healing. But I am seeing now the ways that I'm still bound, paying with my days for someone else's dream, paying with my self-worth in a system that devalues me. Maybe the tight knots tied across your soul rub worse than the ones on your wrists. Part of me says, play the game: how can you optimize your marginalization by manipulating it in your favor? But if I become these metal cogs and pins, I will have lost. I want to stay as love and softness, but I feel the sharp teeth of the gears cutting into me: others taking credit for my work, being interrupted, underestimated, ignored, and cut out. In a culture deeply ingrained with the suppression of everything feminine, youthful, or progressive, how do I be Love here? How can I be compassionate for those who propagate inequality or even misogyny at my expense? Is it my job to teach them? How can I love those who are willing to harm me? Surely the loving thing to do is to reinforce my boundaries, to disengage, and to distance myself. But I dream of work that doesn't require armoring, insisting, and defending.

I am a purist, a scientist, a collaborator, a student of the world, a lover of this Earth and of People. I am not a sales person, a worshiper of money, an ass-kisser, or a climber of invisible ladders to empty successes. How can I opt out of playing without getting run over? There are some so driven to a progress that looks like destruction, so closed when I am open, so willing to take from me and offer nothing- namely one man gunning for my position. Do I fight for *this*? It just drains me. The Wheel and all the interactions within it just seem so valueless, so rigid and inconsequential. How can I add depth to them, or how can I get out before I become the kind of person that cares to play these insidious games of competition?

I vacillate between rage at an entire culture of systemic sexism, anguish for my situational complicity, and disgust for those who perpetrate this death by a thousand cuts. None of them as individuals are malicious- they just move through life propped up by the gender and economic importance bestowed upon them by society. My current contempt is not at all to scale with a day or a week's worth of injustice, it is a dam straining under the load from a lifetime of this.

I want that every time I feel these little violences against my soul, that I become somehow *more* loving, more sharing, more resilient, less tangled in things that do not matter, more open, less like those who are trapped in smallness, image, cutthroat competition, greed, and selfish outcomes. Let me do it despite them not because of them, all love, no bitterness or resentment. Let me be light.

I outgrow my heart again
Tight,
constricting and stretched, until
it cracks apart like the brittle,
thin skin of a locust.

And the new heart, verdant and tender
Wings wet
And yet unfurled
Only now realizes that this
Is birth.
Over and over
I shed my selves
Leaving them in forgotten somewheres
Never understanding the dream
Until I wake,
Exposed and new.

Right now I suspect
this moment may be suspension,
Lucidity:
I feel the glossy boundary of pain
Transforming me again
Always without the respite
of a comforting and silk-spun cocoon,
But rather
with the splaying of keratin seams,
A glass womb with no mother,
The breaking
Of a life that doesn't fit.

I am emergent,
On the cusp of my Self
Pushing to let go completely
And still not knowing if
I will fall apart,
dissolved in the medicine,
or open new eyes from a form
yet unrevealed.

Right. On. Time.

It wasn't twelve hours after lamenting about work, grinding through my thoughts on a situation, lying awake, racked with angst and trying to alchemize it or make an escape plan... out of the clear blue the phone rang, and it was an old professor, offering me a job at an earth science research nonprofit he presides over.

The board has thought highly of me for years after meeting me at various functions, he said. They've been seeking ways to get me involved, he said. Now they have the money to offer me a position- can I consider it? I can work from anywhere, make my own schedule, and the job will mostly entail storytelling to keep interest up and alumni involved.

Oh, Universe, how I love these jaw-dropping synchronicities. Thank you for this gift, for whatever joys and lessons it may bring, and for so immediately reassuring me of my relevance and potential. Onward and upward.

Argentina

My world opens and swallows its own tail, new layers of self and reflection on otherness cyclically assimilating complexity at a hastening pace. A new way is being born into my hands but it's slippery, missing an eye, still shapeshifting as it decides how to manifest. I am different. My interaction with the world is changing, and the familiar patterns beg for new meaning as the kaleidoscope endlessly twists. There is little time to catch up.

A week in Argentina comes and goes in a blur of work meetings, business dinners with colleagues, and long days in the field. But just below the surface, I am metamorphosing. I have always kept myself hidden from others in the industry, believing that my progressive, romantic, unconventional approach to life would further outcast me in the conservative boys' club of my employers. I let more of myself be seen than ever, and though I didn't feel especially held by my coworkers, I didn't feel as ostracized as I might have imagined, and I was elated by a brilliant wine-fueled conversation. Hell or high water, my authenticity increasingly demands that I remove mask after mask... this felt like sticking my toes in the water, wondering how far sharks would go just to eat toes. In the end, I know that I'm not fooling anyone anyway. Masks rarely do. It felt like a relief to not constantly cover my tattoos and my opinions, whether or not anyone judged me any more than before. The only way this position is at all sustainable is if I can be respected as I am. Maybe, I think cautiously, I have underestimated those around me? Or maybe I have reached the point where this job, or any job, is not worth pretending to be someone I am not.

Paris of the South
Raw with concealed delights,
Complex and as
Moody as the sea
Laden with dark romance,
Sweet secrets ripe for the keeping.
An entire empire built from
The dramatic loves that stir the passions of the fates.
We feed them with ours to appease them,
and I break ground on a monument to
a sacrifice
I can't yet understand.

I met your gaze and you said something
I didn't quite catch,
Muse Eyes, you called me
But there was more between the lines,
whispered through a crowd or
held like tongues in a pause,
Slung like the spell of intoxication over the
pulsing metropolis.

I walked past a world suspended in a dream,
Strung like clothing drying on a line,
Like our clothing
strewn carelessly across the floor of the whole city.
I told you not to fall in love
But I break my own heart
until I can fall into the waiting arms
of the Southern Cross.

Clouds drinking up
All that has been spilled
So that there are
No longer boundaries to the
days that became years
And the tears that became oceans,
But only the gray blur of things in transition.

Home

A dear friend came to town and met me for a drink. She proceeded to tell me that my poetry has been her chronicle: her life is in flux, and she feels my journey deeply. She quoted parts, said she rereads certain ones and finds missing pieces of herself in the dogeared pages.

I love her for telling me. But while I am at once satiated with the rare luxury of feeling seen, I ache for her acutely. My work is dismissed or ignored by many, and this doesn't bother me in the least. Those who don't understand me are usually blithe or superficial types leading reliably even-pitched lives. No one without parallel trauma wants to read about the dredging of a soul. I pray so hard for the ones that bleed the words along with me, those who have also pitched unanchored in the vastness, those who yearn for a wholeness that eludes them as it has me. Existentialism makes for an unostentatious kindredship, one forged in the lime-grit furnace of a Knowing. She talks some about how her life has fallen apart, and how my words have been a lodestar in the dark of it all.

She tells me she feels me becoming a goddess, that I've touched otherness. She sees my struggle through my plight as a hero's journey. I tell her she's about to be very disappointed to find out how fucked up I still am. I break through to heights greater than myself only to fall back down into disturbingly predictable mistakes and vices. I'm addicted to attention, escapism, my own suffering. I'm infatuated with anyone who might save me from myself. At my worst, I wallow in self-loathing and objectify my own depravity. I can't pay attention. My ego is ashamed of the dust I stir, wants to justify the parts where I fail myself, the times I make the worst possible choice with zeal. I fight inner wars and both sides are bloodied in their defiance. Gems of love and wisdom are mired within the tangle of 500 other pages of scatterbrained manic nonsense. I tire of my own bullshit. The higher I level up in my consciousness, the deeper of a pit I seem to dig in contrast. My range is expanding, but I am not always winning. I make enormous messes of my life and retain some childish hope that I won't do it again, but when I get bored, I want to fuck and set things on fire. I want to feel something bigger than anything I've ever felt, again and again. I want to put my fingers right inside the snarling maw of wild aliveness, out of some insatiable bewildering pull to destruction as a form of self-expression or penance.

And once they've been sufficiently bitten, I'll desperately crave integrity, purity, and transcendence once more... and the renewal bath of sunlight will bake clean the dry bones of my day-bare self-image. God, how I want to be worthy of the praise she lays across my shoulders. But I am still in the boiling pot, scalded by my own becoming. I call abundance to me through the ethers only to trip over it when it arrives, unable to feel worthy of the very things that would nurture me. I am trying so hard to save my own life, but admiration is grossly inappropriate for something like survival. The medicine is as ugly as the disease. The pendulum swings back and forth: enlightened, embroiled. And I apologize for neither. But I despair from the growth of the polarized parts of my psyche simultaneously expanding in opposition. I am not yet well. I am not ashamed of clambering for grace through the wreckage, but surely the graceful don't succumb to such callow longing.

Yes, I tell her, I have touched the faces of the gods in dreams, but I keep the other hand blistering within the furnace of my heart that burns through this material life for sensory fuel. Some days, the wrong wolf wins, starving and carnal, and she, too, is me.

She quietly lets me finish my self-depreciating rejection of her flattery, and is neither surprised nor quelled in her veneration. My heart twists, and I hate it all the more for her that she understands completely.

Costa Rica

Dawn kisses the frozen horizon
And I am at once
Above and within the world
At one with nature and yet outside of it
At one with love
And still unable to allow it.

Fishing around for loose change,
Lost treasures,
Dropped in the
Cracks of my soul.

It stays split apart just enough to
Look into
Or reach into
But never both at once.

Dream:

I was in Cape looking south near the interstate, and there were tornadoes on the horizon- then quickly they were really close to where I was. I commented that it was the closest I'd ever been to them in real life, but as I said it, I realized this wasn't real life but was a dream. Other people around me continued talking and bustling about, concerned and interested in the tornadoes. I kept saying it doesn't matter, this is a dream, wake up, wake up... but I couldn't wake myself up. I was trapped and lucid. The tornadoes changed into giant female forms, like gray genies or goddesses. Then they changed into or created huge gray singular breasts, that were mounds on the ground. I remember men (from my life) climbing on them, squeezing and sucking milk from their nipples. Erotic, self-serving, removed, nourishing... complicated emotions overlapped. I was the breasts, and was on the breasts. I desperately needed sex and nourishment. Then I was at once the hunger and the food, the need and the fullness of climax. I was both this terrifying female energy and everything that uses her.

The unconformity of you
Represents more in its emptiness
Than could ever be recorded in the
Millions of memories I keep on either side of you

You
Eroded everything that I already was
And now you exist only as
A lack
An absence
A nothing
In the layered story of how I lie here,
Existing.

On Icarus wings I remember you
One last time, up close
I use all my last love for you
To burn up completely.

It was worth it.

Home

My Medicine People call to help me. They feel that I am undergoing a transition into embodying my power. I take them very seriously, trying to keep my dumbfounded ego at bay. Power feels like a word reserved for anyone else. It is unusual for us to speak so directly about the energies of our lives this way, but the resonance of our conversation is undeniable, and it seems to rattle some long lost truth loose within me. The challenges with my mental health, the lucid dreams, they tell me, are all symptoms of my awakening. It is soothing that they not only *see* me, but they *hold* me through these last few years of madness. Although we don't talk often, they are among my dearest friends- at once parents, siblings, peers, and mentors. I look at their beautiful faces on my laptop: she, glowing with some rapturous inner grace- an understated, nurturing, but impossibly fierce resolve. He is a lion, a hyena, a shadow man playing dice. He is dark but good, a conduit of transformation, a bridge between worlds. His presence commands some kind of cellular regard, a subconscious genuflection as everything around him is entrained into his gravity. He is a death doula, a seer. She is a protector, a guardian, a true healer. They are yin and yang, Shakti and Shiva, the moon and the sun. Just being around them moves me.

He speaks to me in an encryption that protects him from the accountability of directness or steering me into the depth of my journey sooner than I am ready. I understand him, I think. It's as though he is transmitting a message not entirely from his own perception, and I listen from beyond my hearing. I quote him, but it's paraphrasing. He can't be pinned down; it's part of his mystery and a result of his wisdom in not taking on any part of my challenges. I am dying to ask pointed questions but I can hear him in my head, feel him. I know I cannot ask, that I will find the answers, that they are not his to give. And anyway, they will be answered by trials that must be endured to be understood. I am green, but there are some things I know without having been taught. Right now, I'm basking in the warmth of his attention, his advice, lapping it up like milk. It is rare and extraordinary when my esteemed kindreds offer this kind of direct insight, and I overflow with excitement, validation, and gratitude.

"Don't apply labels," he says. "We don't yet know what you are becoming, but it is *something*, you are clearly on an initiatory path... these trials are capable of making or breaking you. Arriving at your own healing is a matter of life and death. You are sensitive, and of course you can feel some of these things already.

"I can only tell you so much, the rest you will come to know. It is good that you are learning many different Ways in your travels. Traveling is clearly in your medicine." His eyes flash like lion eyes, the way they do when his words have double meaning. My travels will not always be on the physical plane.

"Your caliber of lucid dreaming is a sign that you are transitioning into the next level, you have graduated one school and are beginning another. The fact that you are becoming trapped in your lucid dreams is a warning that you are not in control. Your powers are growing more quickly than your ability to employ them properly. You have to get a handle on this, or you risk going and not coming back. The dreams are real places, and you need to protect yourself or it is impossibly dangerous to be there. Learn to look at your hands. Bring them up in front of your face without them flickering.

"Create a suit if armor for yourself, whatever it is, make it elaborate. It must be made by you and it must cover you completely. Protecting yourself is absolutely paramount right now.

"The goddesses and beings you see in these places are real forces, real entities- archetypical energies manifested. When you become more adept, it is clear you will work from within your dreams. You can travel to these places to get what you need to heal, but you must recrystallize with perfect integrity.

"You will learn to make pets of your shadows and they will help guide you. It's time for you to seek a teacher, you need some mean old woman to teach you the traditions. For you, a man *cannot* be your instructor."

We all laugh. My love addiction has created some legendary dramas in this life. My power struggles that play out with men of power- or even arrogant men disguised as powerful- is not news to anyone who knows me well. Ah, the pitfalls of a passionate heart.

He continues, "It is not just you, honey, shaman men are opportunistic. We are still human after all. Seek an old woman, she is already looking for you. You will have to withstand her sending you away and humiliating you. This is part of the learning. There is no mystery in the ways she will eventually teach you. The traditions are to-the-point about these other worlds. But you need guidance and formal training. You are getting stronger, be careful what you wish for. The more powerful you are, the greater tests you will face, you are capable of drawing to yourself a proper enemy now."

* * *

Later, I write them:

I did some work last night.

I consulted the cards, and I could hear my people howling in laughter: they said I can't read the cards because I don't need them. And still they won't just tell me what it is I *am* to do, what it is that I *do* need.

I realize I have been avoiding feeding Spirit with proper ritual because I have been afraid that my ego will become involved, that the excitement from feeling agency at all after many years of powerlessness will intoxicate me. So now I ask how to sacrifice, how to listen. The sacrifices I know are severe and amateur, and I don't think they are the right ones now.

Thank you for sharing your wisdom, I feel it coursing through me in its life-giving acuity. Though I don't entirely know what is unfolding for me now, it feels incredible to have people in my life that believe in me to help guide me through. I am thinking a lot about falling apart and coming back together, and the art of doing this with intention. We do this over and over with our emotions, our narratives, our patterns. I had to live through being completely shattered apart and recrystallizing in the waking world to arrive at some knowing of how to do this in dreams- consciousness learning to become aware of itself on both levels. I have always been fascinated with dream interpretation, and I have only recently realized that we live out these stories and dramas on the top side as a reflection of what we need to level up... the subtle is in parallel with the gross. Our dreams reveal our subconscious landscapes, but reciprocally our waking life reveals what healing we can seek in etheric lucidity. We play out our embodied myths in the physical world, the ones that were written for us in the Everything as leverage for our lessons. It's all a symbolic mirror.

There are so many names that you have, but sometimes I hear the real ones and I want to sing them back to you, back to the world. I know that my teacher is already sending for me too, and I am awaiting the moment of intersection. *Are you sure it wasn't last night?*

The physical alchemy that the soul comes here to undergo cannot be understood beforehand. The process is necessary, but messy, imperfect, directional: chemical reactions that forever alter the state of being. The soul is completely surrendered to its embodied form in the most beautiful destruction, as natural as it is cataclysmic, as loving and simultaneously painful as birth... we come here to experience the poetry of metaphors, to live out the prophecies of dreams that we vaguely remember from our time in the Knowing. We come here to feel the cut of the knife, the devouring of form by the fire. We come here to feel the paradox in the duality that is love, and ultimately, to be consumed by it.

Surrender

Sur-render: *Sur* meaning "over" plus *render* meaning "give back".

Surrender- to give oneself over.

Houston

How much are we really asleep? *This* is the side where we don't understand. This material plane where we feel pain and confusion and the weight of our bodies, where we are enslaved to the biology of ourselves unless we dedicate our lives to developing our consciousness sufficiently to overcome it. This is "asleep". When we wake up, we are light and free of these burdens, resorbed into knowing and love and connectedness.

Cancer.

I'm at a conference amid a bustling cacophony of people when I get the call. The word punches me squarely in the gut, and everything around me is slammed into a dizzying amalgamation of handshake-pleasantry-smile-laughter. I can't hear properly and can't escape the vortex of business superficiality fast enough. My senses swirl in the sea of dry-cleaned monotones and their corresponding sales pitch faces. I run outside to catch my breath and a cab to anywhere else.

My hands ply atop pale soil,
Hard earth that won't receive me,
Won't open this time.
Burrowing like stubborn roots,
Grow me here, I beg.

I want to stay now.

But nothing is promised,
Nothing is promised at all.

We are carved into the bark of oaks
and swept away like loess,
Countlessly erased and accumulated;
Effigies of us collected or forgotten
We are treasured or lost in obscure caches-
We are gods or ants,
illiterate and oblivious;
We consign the spent tokens of ourselves:
shells we outgrew and left on the sea floor,
proof that we loved, flung into the wind
to drift like feathered seeds.
We touched things, we changed things,
rearranged furniture,
Our eye teeth fell into the medicine pouch,
We lost blood in the woods, running and wounded,
We lived.

We are cherished or left to rust in
dimensions we can't access,
Swan song playing softly,
We wind around our lives in silk knots,
a cocoon we make for the legacy we cannot remain within,
Standing still and unraveling
as the earth and everyone on it spin outside of us, pulling.

Let me never need to know
The ways I am kept or abandoned,
Let me stay, I beg.

Let me be written once more.

My fingers are mangrove roots
Splayed wide to hang on,
But still it comes.

We are the last to let loose the grip we have
on the story of ourselves
And leave our pages on top of the earth to
blow around as they will
Relinquishing to our scattered owners the
onus of returning us
Unbound,
Unfinished.

Home

My body is diseased, and I am afraid of its death. A death I have often idealized in the throes of my depression- but in noble ways, not in some graceless deterioration punctuated by glaring fluorescent lights and the ceaseless amazement at the threshold of human pain. I wanted, and still want, a Real death when it comes. Something that feels like readiness. Ritual. A clear path over.

Everything about my diagnosis feels so *clinical*, all facts and measurements, no humanity. I am biological data that is carved up and fed into machines and microscope slides. We know very little so far- there are no anchors of reference for my emotions to contextualize what will happen to me yet. Is this just a little bump in the road... or is this *it*? The word cancer feels like a prison sentence, and I am waiting to see if I'll make bail. The range of possibilities yell to hear their own echoes the dark halls of my mind. Today's exams were invasive: blood work, scopes, biopsies. I am the wounded feminine, the shroud of consent foisted upon me and inferred by my presence, my consciousness detaching from my body as it is painfully plied. I do want clarity on my condition, but *is this what a healing process really looks like?* It feels more like hell than help.

And now all there is is waiting. And worrying. And wondering if I have caused this for myself with my behaviors. An inclination to bargain with the universe arrives, and I indulge it. I have moments where I pity myself immensely. Even if this doesn't kill me, I know that the marathon of pokes and probes and waiting for answers has only begun.

I sit here wondering if I have 5 weeks or 50 years left to sort my shit out. Isn't it strange we adapt to our perception of that? I haven't finished nearly what I came here to do. I am not sure I've even started. And I am pissed about the time I wasted trying to please people.

I feel so unavailing and remorseful that if I were to die, the most important person in my life may one day read all these tomes of drivel and barely see his name. My heart is nothing but pages of letters to him, but my journals seem to be everything else. How can I tell him he's been too sacred to write? Could he ever understand that confinement into a narrative is a violence I try to spare him? That what I love about us is our continual escape from the clutches of definition?

I want to go home to him, to the times before we were so complicated. I want everything else to fall away so that I can crawl into his arms, our love, our home. I keep trying to fix myself before I allow myself close to him again, to protect him. But I've been gone for *so long* in one way or another. What if I'm out of time?

Dream:

There was an animal like a bull being constrained by men. One of his legs was missing, and he had giant horns that came out from the full diameter of his sometimes human-looking face. We were referring to him as a Wild Beast as though it was a common animal name. The men wanted to ride the wild beast and kept trying to climb on it. He was bound by ropes from multiple directions. I remember watching and not participating. I felt sorry for the beast. Later, in a kitchen, there was a discussion of the plan to ride him- it was so important to them that they accomplish this. I was wearing long skirts and had very long hair. I was leaning against a counter. I felt no intimidation nor sympathy, I felt very little for the people in the room. I said, "I am a witch, and I could use my magic to ride him- but I can't use my magic for you to ride him." I remember I had no interest in riding him at all. The dream changed and I remember seeing a commercial, they had made the wild beast into french fries.

Two years ago today I was being broken open into newness by my first solo travel experience. Now, so many lives later... life is finding new ways to coax or break me open.

Once you begin down this path, you have chosen. You cannot unsee, unfeel; you cannot reject your growth or your knowing. There is no stagnation or regression back into the naïvety of ignorance, there is no pause. You learn to draw breath with reverence and gratitude. You forget, and are shown again what is required. For every time you have loved, you will find how love can be ever more powerful; for every time you think you've been cut, you will be split apart more deeply; for every time you think you know how to let go, a bigger Letting Go will find you to teach you.

Every time you get lost, though, you will know more and more how to sit with your inner wayfinding. You come to trust yourself and the unknown as life plays out its mythologies, the unveiling of mysteries and deeper wisdoms; the recognition of mortality and continuance and how they apply to you. You learn to witness your own reactions and to hold yourself compassionately despite your humanness. Even as you believe you are getting good at knowing where you are, there will still be times you get lost. But we can learn the portents, we can understand and make peace with the process. We are not being conquered by our trials, we are being transmuted.

The way out is the way in.

It's right on cue if this disease will usher me into my next level. It's part of the initiation and I feel it. It's a test but I am not entirely sure how to pass. If I can heal myself now, I will surely be new.

I sit with this, alone, for days now. Andy has gone to visit our families. I haven't left the house. Everything I am doing, my entire way of living, is ritual. Self-care ritual. Bandaging ritual. Visualizing the cells restoring to pink and healthy. Essential oils, medical journals, spiritual coincidences of helping other people. I am in the pause, only listening, and so much is there to hear. I forget that I should just trust my gifts and stop explaining them. The scientist in me fights, justifies. But how do I know the weird things I do sometimes? It's not remembering, and no one told me. The healer part of me just Knows. It's inexplicable, clairsentience maybe: I spontaneously see things, feel things. I am tapping into subtle information, and it happens more frequently again now that I am quiet. It reminds me that I could feel these things when I was a child too, but it got lost somewhere along the way, I shut it off when I stopped trusting myself. I can look at people and know how they are ill and what their challenges are just by the energy of their bodies, some *way* that they present. I can also feel elements and animals, essences around them. I am not sure what that part means exactly. One friend is rose quartz; one is orchids; one is wood but also a shapeshifter, some kind of weasel usually. My love is clouds and swords, green-grays, storms because of the wind.

What am I? I don't know exactly, that's not as easy to see. Sometimes metal- silver, sharp like little blades, thin like feathers, but smooth and flexible like snake scales.

I can feel what I need for my healing- onyx to absorb the negative energy; eating purple things that grow from the earth; quietness, cocooning; deep sleep, forgiveness. Dreams, the Coat of Armor. Frankincense. Palo santo. Love. I am being transformed, and I have to be sure I get put back together correctly- it is my job this time. I can direct it now, I am not falling all the way apart, but there is a responsibility. Boundaries. Creation. Selfhood. Who I end up on the other side this time will be the most Me I have ever been. It is a time to listen deeply and a time to be deliberate. There is no room for error. Thoughts float in and out, and I let them through without judgment no matter how odd they seem. I am in a trance.

I want to heal dynamically, using both sides of myself. But there are times I think I will have to leave the scientist behind. The healer is permitted in places that the scientist is not. But the scientist insists on understanding. I think that is ok as long as her need for control doesn't interfere with the necessary deconstruction.

I am processing a burden of ownership over my illness, that it was self-imposed from my own actions or potentially my carelessness. I accept the consequences for living inside this body, for where we have been and what we have done, but I challenge this ownership of illness at the line of feeling guilt. I feel compassion and empathy for my selves who made our past decisions... we did the best we could to survive, to feel loved, to retain health or sanity, to get by. I refuse to surrender into the slavery of anyone else's perceptions of whether my decisions caused this fate. If this is mine to bear, then so be it, I will bear it. But I wont be strung up as someone else's example. Having a body just has consequences sometimes.

Because of the nature of my cancer, it is hard to be as open as I might be otherwise. This, too, is part of the lesson it seems. People around me are being sorted into tiers, culled. I focus on allowing myself to be held in the care of those that show up. The support of the few closest to me- those who understand the test that is playing out- have been incredible. No matter what happens, I just wish for grace. Andy has been warm, nonjudgmental, present, comforting. The Lion and the

Guardian work in the background of my healing, I feel them. They hold me steady so that I can do the work.

I am now my own mother, that is complete and obvious. I don't spend time mourning for what I cannot have. I stand here and nurse my animal back into health, and I am reassured and nurtured by this side of my self. I am steadfast and permanent, the exact mother I need. I can't tell my Dad what I am going through, for a few reasons. Mostly because his worry will inadvertently throw my energy off, and I will be back to taking care of him instead of taking care of me. I have become both of my parents to myself, and for vastly differing reasons.

Dream:

I am in a doctors' office for a surgery. The room is like a cubicle in an office. I am being hooked to IVs. It's a big procedure and I am afraid. I wake up after the operation, I'm in terrible pain. I have had "bone surgery" and many of my joints have been operated upon. My left shoulder has been replaced with bone from my right hip. I have bandages and sometimes where they are falling away, I have open wounds from the laparoscopy.

Dad and Kimmy take me to Grandma and Grandpa's but they aren't home. I vaguely remember that they died, but I am in and out of consciousness from pain and I can't fully perceive this. Beth Ann is there. She sits on the couch by me and says something, tells a joke and then absentmindedly smacks my shoulder as she is laughing. It's excruciating.

All I want to do is sleep. I have a backpack with me, and I take it into the wood stove room and start to make a pallet to lie down. It's daytime and not time for sleeping. There is the feeling that I'm not supposed to be here, that I was supposed to go home instead, but I am relieved to be someplace familiar and safe- I just need rest. Wherever I am supposed to be will wait, all can wait.

We are all back in the living room. Grandma and Grandpa are home now. Grandma is sitting beside the Christmas tree in the floor amid a pile of wrapped presents. She says even though it's a few days before Christmas we can open presents early. I remember wishing we could wait to open presents until after I had rested, cleaned up, and changed my surgical bandages.

...

My gifts are being given to me, ready or not. It's time.

Honoring the gravity of an illness is required to cure it.

I have been in the pause, ear pressed to the earth. I have been in my own presence, tending. The Lion says that cancer is what happens when cells no longer remember to what they belong. Ok, I think. Then I will heal my cancer by embodying my own belonging. This feels right. I will step forward into my Self and own the world I have created, allow myself to take up the space that I need, allow myself the importance that then permits things to belong with me. I will stop abandoning myself and reclaim my body as my own. I will retrain my cells that they are part of this Me, that they are vital and loved. This nurturing is where the miracle lives.

The scalpel tells the body: *these parts are no longer You, they have betrayed you.* The song tells the body: *remember, remember, re-member, my beloved.* Maybe there is a time and place for both, but one is violence and one is love.

This is systemic- the cancer is a symptom, the same as my mental afflictions. It's all part of the same imbalance. If I can return to resonance with my own belonging, maybe my mind and body will follow.

Baja Sur

Glittering coals of a smoldering city
snuffed by the dense wool blanket as
I rise to meet what awaits me
Above this life
And back out in the wild.
Higher still,
Ten thousand eyes watch patiently from
the blue forest of my ancestors;
They will me to walk the path they've lain before me.
My feet set out before
my mind can grasp the assignment
But too late,
The fates delight and punch the ticket
unfurling now the mythology and
narrating the script from the peripheral,
Eager puppeteers lip syncing the lines of
their favorite play.
The strings dance me,
So I dance.
They pull me and twist me and
teach me the movements to manifest what I need.
And thus I recreate myself
Again and again, in a theater of one
Surrendering to the silver thread,
Allowing this to flow through me.
I'm relinquished to stay the course
For as long as it takes to embody the story,
The archetypes guiding me to
whatever may end me
or complete the alchemical medicine to finally heal.

I thought I knew what test I came here to face, but now I'm not sure that I understand. Yesterday I checked into to the hotel, had a margarita, went to the pool. My friends arrived and came swimming, and we drank and laughed until the sun slipped westward, casting long palm tree shadows on the water.

I guess I had too many drinks, but my subconscious balks in vigilant rejection of that as a full explanation for what I experienced. A gap in time and space opened up to swallow me, and while the margaritas perhaps increased my spiritual porosity, they certainly were not capable of pulling my mind apart.

Returning to my room, dripping wet and giggling, I realized I couldn't find my keys. I looked up and noticed that my door was oddly unlocked and ajar, and I was immediately sobered and unnerved. Thinking I'd been robbed, I called for my friends to come inside with me. I found my keys on the living room table. An orange bag from home that I *know* didn't bring with me was in the chair. My mind was spinning- what kind of cosmic fuckery was this? It was more than bizarre, and a wave of fear enveloped me, prickling at the back of my neck. I felt like I was losing my grip on reality for a minute. I tried to explain it away: surely I brought the bag and had forgotten. Surely I'm the one who left my keys there and the door open. Surely being in the pool and having a few drinks had bent my mind a little. But I've been saturated with sun and tequila many, many times in my life, and I know where that ride is capable of taking me and and where it is not. I am high-functioning in my indulgence, propped up by my tolerance and heritage. As inconsequential as the anomalies were, they were disturbing glitches in the matrix, Easter eggs discovered in a dissonant movie.

My friends are not the types with whom I can discuss a spiritual crisis. Lovely as they are, they are atheists and pragmatists. They forgave me for being dramatic, and we shrugged it off as me just getting too drunk. But this morning at breakfast, they admitted that their tv came on randomly in the middle of the night. A lot of energy is moving here and I was caught completely unprepared.

The more I feel into what happened, the more I sense some kind of darkness trying to undermine me, eroding my clarity and letting my mind play tricks on me. There have been multiple instances of items going *entirely* missing at home lately, and odd things moved around in the house. Then I come here and find something of mine that I know I didn't take with me. I hate feeling like I'm losing my mind, but it is *possible* that I just don't remember what I do, what I bring, where I put things. Alternatively, something perhaps wants me distracted, upended with confoundment. In any case, I can't drink or let my guard down again while I am here. I don't want to talk to anyone about what I am going through in case they might think I've gone crazy, or worse, they have some kind of hidden or energetic agenda against me- whether or not they know of it. I feel vulnerable and exposed. Ugh, I reread this and it sounds paranoid and insane. Am I becoming schizophrenic? Can I still trust my intuition when it gives me information that cannot be rationally explained?

I was worried before about making it through this trip gracefully and deftly, energetically finessing the narrative as I navigated ending a complicated relationship. Now I am worried about making it through this at all. I feel spiritually sick or attacked, but I don't even know if that is a real thing.

We waited so long for these moments and now they pass stupidly, empty of content, a slow swirl around the drain. You offer nothing of substance: no words, no wisdoms, no laments. You seem to have nothing left of yourself compared to the You I knew from before; you have allowed your horizon to be erased, your boundaries to be drawn by another. You're condensed, simplified, serious. The ways that you were playfully twisted have become hardened and severe, and though we just spoke last week, a stranger stands before me. I can't save you, and you don't ask me to now that we're in person. There are so many things I want to tell you and ask you, unlived stories between us pushing me to pose questions with no answers.

I'm sick with the spiritual gravity of circling around this. I'm past the point of being able to hear. Sentences tangle, time runs together, overlapping in fast-forward plastic ribbon cassette jams. Why can't I say what I came to say? No angels rush in to carry me, no dreams bring me medicine, no grace or vitality arrives. I have no hands, no teeth, no presence.

I want this to end, but I am horrified that it drags and sticks, shrugs, stabs, avoids. *Not like this.* I can't hear my thoughts. I can't remember songs. I'm dizzy, quiet, bloodshot. Days go by in a blur devoid of potency. I have a knot in my gut and can't eat. It ends anyway, but with my mood since we don't talk. It ends with my awkwardness and angst and confused entanglement. I revert back to the games of cowards, and I make you end it because I can't bear to hurt you even when you're awful. I sulk because I am disappointed in my regression. I want a clean cut, a sutured closure, but I've botched the operation. I am surprised and empathetic when you lie to me because it's so absurdly unnecessary. You were so grand in my mind, but here you are: small, devolved, anemic. I comfort you for your shortcomings, and inside I die a little at my ridiculous sacrifices.

I have to write you quickly, encapsulate you in kindness before I process the sting of your crimes against me, small and large, subtle and outright.

I mourn the construct of you but not the man, and sullenly resign to the recognition of a familiar plot: idealization and disappointment. *Give me something besides my delusions to miss. Give me an honorable ending. Give me a worthy grief.* But time passes, slow and glaring in its reveal that we will not give this a proper burial.

You grasp for tatters of the fiction that now shreds itself against the sharp, angular reality of us in the flesh. We are different than the idols we built and worshipped, now flawed and grotesque in the sunlight. When your hands can't catch the pages spilling from the broken binding, they find my throat again. With the fantasy dead, it's nothing more than an act of violence this time. Are we discovering this together or are you punishing me for your emotional impotence? I feel empty, sorry for both of us. There was a less messy way to do this, but apparently we had to destroy this as savagely as we created it for it to really be done.

What can you give someone to take their power away? A kindness in the face of their cruelty, compassion in the contrast of their judgment, a graceful surrender to a life unfolding with maladroit endings. You probably think I'm a fool, but as long as I still love, I win. I wonder if life will ever open you back up. I wonder if you'll ever look back to reflect on this with some newfound gentleness. I want to leave this lovingly as I leave it completely, knowing that it's really the only powerful thing that I can do. I feel not just the loss of you, but an immense sadness about the brutal world that has inured you in kind.

The last day, you offer a photo to me because I want one. I want to remember you outside the context of this discordant week-long goodbye. I want to remember you how I did before I saw you again. We fall from the pedestals together, separately. How differently I see you now. I see the shadow of you, black like oil floating on the water around you. You are an unapologetic darkness that has moved in me like medicine or poison, depending on the dose. But where is that part of you standing solidly to interrupt the sun, surely bathed in gold? Where are your edges now? You are dissolving and formless. You are liquid, spilled.

In wrenching contrast to this paradise, secret hells eddy in the periphery waiting to pull me under if I don't keep paddling. I know I didn't see those last time.

Nothing is meant to last forever, and circling back to places to witness them with new eyes has given me a revelation of bittersweet understanding. In our transience, we collect only moments. The shutter snaps and the feeling of a time and place, a love, is encapsulated. But feeling something and *knowing* something, someone, somewhere, are worlds apart. Feelings are always valid, but they may not be based in a context that transcends.

I cared for you without knowing you, as I have with so many. Not a big, life-upheaving love, but a friend love, a love that hurts to lose, nonetheless. I can never regret loving in whatever form it takes... this, too, has a place. But that place is not made of the same things that make a home, and home is what I crave now. Loneliness can't be palliated for long by drifting about insolubly on top of little loves. True inclusion requires that you mix, you assimilate... that you Stay.

For all the times I have longed for someplace to belong, for a *home*, I haven't yet tried to sit still and make one. I needed to heal and that led me to places, to people, who evoked novelty and adventure and raw aliveness. And now, that perhaps has run its course. Now, there is nothing I want more than to know something cellularly. I want to grow into a place, a love, and have it receive me as integral, necessary. And I want to let those things grow into me, a sharing of responsibility; a seeing of the secret faces only revealed one by one, slowly, requiring stillness and geological patience to see and learn.

It takes a broad kind of time to belong yourself somewhere and know it inside out. There are no more loose ends anywhere, no more ways to escape and no more reasons to try. I've been running so hard from my own life. I think it's time to go home.

Home

I'm going to belong myself to these mountains
and they will feel me
like they feel other wild things,
Necessary and familial.
I'm going to re-member myself back into the loams
and crystals
Roots for fingers
Slowly digging to connect all of me,
Incorporate me;
Lungs like branches,
spreading, splaying,
Steadily breathing me into the above.
Each of my pieces finding its origin,
The sum of my existence finding a place to rest.

I have learned that homes are made
And not found:
A contract to neither cling nor cleave;
Sitting inside the sameness,
the reciprocity,
I know that
These places also belong themselves to me.
I am alone but not separate,
Integral but indistinguishable,
Born from this land
for the sole reason of returning to her,
A journey of inevitability
that can be savored but not denied.
I am a mirror, pondering.
I am a self, Selving.

There's a ringing in my ears
Like wine dipped fingers
Circling the crystal

You are remembering me
Rinsed clean after the forgiving

There's a ringing
Like tiny brass bells shattering the
Parts of the story that used to stick in our throats

The quietest symphony ushers us now,
Delivers us back to ourselves
Whole again.

Dad's dream:

We were at a really nice place, like a resort but that is not the right word, I just remember the feeling of comfort. You turned and smiled at me, and then all of a sudden there was this horse- chestnut with some white, with beautifully detailed luxurious tack... you smiled at me, and then you got on this horse and took off. And I remember thinking, I didn't know you could ride like that. I didn't know where you were going but it felt like you were going somewhere good- not running from something but rather toward it. I saw you take off across the side of a mountain slope, about halfway up on a huge, like 8000-foot, mountain... For a minute, you dipped down on the side, but then you came back up and if anything you were going even faster, like the wind. And then, you were gone.

And I woke up feeling so happy. The dream was so real that I felt like I had been hanging out with you.

The weirdest thing- I just went into the bedroom, and there on the dresser I keep a picture of you when you were 3 years old. I've always loved that photo- it's you pulling a little wagon behind you like, "Alright, I'm taking off now." Anyway, the strangest thing, that picture is turned on its side, just sitting there at a 90 degree angle. I know I didn't move it and no one was in here.

I went outside tonight, and my owls were talking to me. I always feel more connected when I hear them out there. They are messengers. I don't know if the message is good or bad, but they just come to tell me to listen.

Attachment keeps us from exploring ourselves fully by directing our attention disproportionately outward. It's an anxious emotional accounting of that which we have and don't have. We invest in relationships with the hope that we will find inclusion. We pursue material ventures with the desire to feel worthy or successful. *But feelings are generated internally.* To incorporate belonging or worth into our Being, one must access these things organically: healing and growing one's self-perception is the only way. Many people never see that chasing externalities is a losing game- as soon as we have what we want, there is more to want; though we may get what we longed for, we remain yet unfulfilled by the absence of some abstract wholeness we believed would arrive. So we chase the next thing or ache for our losses... when we should be questioning the conditioning that makes us feel incomplete in the first place.

If someone grows up with their physical and emotional needs met and a secure attachment to their caregivers, they are free to develop their independence and identity- the child can stand on a foundation of security and self-worth to safely expand their world. Alternatively if these needs are unmet at an early age, a subconscious disquiet about survival or being abandoned may follow a person into adulthood, manifesting as materialism, fixation, clinginess, or an unwillingness to connect deeply. The dysfunction is transferred and compounded. Healing of the inner child, then, is key to creating the self-trust that gives someone the agency to create healthy boundaries and feel "enoughness". This sense of safety is fundamental to right relationship with the self, with others, with work, and with the environment. Establishing secure attachment in adulthood requires braving an uncharted inner wilderness as a pair of selves: the fearful but curious child and the inexperienced but committed self-parent. As a person starts to consistently provide and show up for themselves, self-reliance becomes a system of steadfast constellations, offering guidance along a lifepath of resolve and integrity that are not dependent on circumstance.

"You are powerful enough now to draw to yourself a proper enemy."

The words echo in my memory. My imagination conjured scorned and jealous women or great sorcerers who want to steal my light, a battle clearly defined in its measure and means. Yet it should be no surprise to me that my reality generates its tropes and tales from the material it has available. The characters in this epic are shrouded in mundanity even as they are formidable in the myths they deliver. My greatest adversaries thus far, those antagonists whose dark gravity spurred a resistance, have started as my dearest friends and lovers, arriving into my saga within the Trojan Horse of beautiful intentions.

I have been so careless and unintentional, bouncing off of my life instead of making decisions. When some new romantic tragedy arrived with a bouquet of red flags, I threw caution into the wind like it was confetti at a grand party. I cannonballed into the tempestuous arms of people who came to hate me when they couldn't control me. I mistook the depth of my connection with soulmate friends who blew away in the breeze. I have entered into relationship so casually free of deliberation that I never stopped to consider what aspect of me was offered for connection. My emotional wounds drug me around the world, seeking the relief of someone else to heal them. But when a person is not conscious in relationship, they are capable of playing out repressed patterns: the familiar push-pull, an addiction to feeling too much, the creation of dramas to enact our unfinished stories. In truth, in my naïvety and unawareness, in my hollow self-worth, I have been my own worst enemy for far too long. I ache from setting myself on fire, from giving away my power. I am learning to contain myself. If a worthy opponent arrives now, let me see them for what they are.

JEN BARANOVIC

Ecuador

Time drips off the clockface
Spilling from the edges
Muddying the floor
We squish our toes in the Now.
We laugh at the circles they made-
Wheels rolling toward nowhere
In attempts to contain and define things
Like moments.

We cup our hands to fill them,
We hold our hands to share them.

This is how we Measure.

The hands of the clock can never touch us here.

At a reef-enclosed bay on Isabela after the first boat left and the sun was not yet oppressive, I stared down a fear of mine that I've had since I was a kid: jumping into deep water. As ridiculous as it may seem, all of my blissfully sublime ocean adventures lately come with the building hidden weight of inevitably facing this. I'm the one who doesn't jump off the rock into the swimming hole, the one who uses a life jacket while snorkeling, the one whose best swimming move is a flailing backfloat.

Today, all I had was a bikini and a snorkel mask, and the grit determination that comes from wanting *in*. I looked around at the sparse few people who joined us in this bizarrely unpopulated wild paradise. *Ok,* I thought, resigned. *I will either drown in front of these people or I will swim.* It felt like even split odds. Nervously curling my toes against the smooth wood of the bottom step on the deck, I strapped my mask on, took a deep breath, and leapt.

The bare and childlike bravery of that timid jump into the crystalline depth defies description to those who haven't yet met a threshold moment. After a few panicked gasps through the mask, I realized that *I was doing this!* I was elated for taking such a big step in overcoming my personal hurdle, and overwhelmed with the payoff: microcosms of dazzling hypercolor life teeming on the corals. I have never seen a reef so healthy and vibrant. The longer I hovered to observe, the more that was revealed- communities of creatures in their dynamic interplay of rituals.

Then... out of nowhere, a sea lion pup zoomed by me at lightning speed! He playfully darted around me, so close I had to stop myself from swimming into him! In a flash he was gone, my heart left pounding in my water-muffled ears from the encounter. I will *never* ever forget this experience, one of the most magical in my life. I am broken open in gratitude that I managed that leap.

I already knew how to swim, I just had to let go and trust it.

Home

The dead don't stay buried for long anymore. Three unearth themselves in as many months, shaking free of the feeble encapsulation of storylines that killed them off in my movie. The tombs of unspoken words that I mentally contained them in could do nothing to stop the phone from ringing out of the clear blue. I stare now at the device, lying there innocuously on the table as though it's not a portal for my baggage. I can *know* so succinctly that I want a more grounded way of life, that I am ready for the gifts of depth and real belonging, but I am not yet free of the wheel of karma that I sent spinning as hard as I could crank it. I haven't yet passed the tests that will refine the life around me. It's easy to believe you have integrity if it is never challenged.

I sigh at the buzzing phone, as though he lives inside of it. He lives there more than in my real life, I shrug. Mexico feels so far from my Tuesday leftover lunch, so far from the weeds I need to pull, so far from everything I can see or hold except that phone. There is no integration of my lives- there is the one here, where I am Real, and there are the memories of being there. I used to think they'd intersect, I used to hope for it. Now it feels like an arm reaches into my kitchen, reaches into my life for me, trespassing into my thoughts whether I pick up or not.

I can no sooner write people than I have to erase them apparently, scratch out parts I thought I knew. No... people are more like drawings or effigies you are continually creating in their image, but you only have the media around you, and it changes all the time. I missed you and I made you out of shells, small beautiful found things, something that remained of something else long gone. You hurt me and I smashed you into sand, let the water carry you where it wanted. You call me and my hands are full of clay, full of your asking: *make something with me, touch me again, tell me how you see me.*

You call through Skype since you deleted my number. Clever. I put down the soapy dish I'm washing and dry my hands, staring at your name on the screen. To answer: consent. Consent for your explanation, approval of you calling me this way. I shake my head and take the call, curiosity overcoming me. God dammit.

You're sorry you were an asshole, you got in over your head. You can't go on without talking to me. You love me, and you need me in your life, you don't know how, it's all a mess.

I let your words pour over me, washing away how badly you treated me, but I keep my mouth closed, listening, careful not to drink you in. I forgive you completely, and I understand more than you could know, but that doesn't mean I want to do *this* with you anymore. This, whatever-this-is that we do. I don't want complicated friendships anymore. Between the lines: you want to salvage the fantasy. Maybe we should have left it there, never tried to take it to the surface. I miss the daydreams too. But the longing is empty, the spell is broken. It's too much to ask. I love you too, and I'm glad you called. It's a softer landing, an ending that I'm glad to rewrite. I make the clay into a nest and let it hold me. We can't undo the whole thing but we can undo the last part. It feels better to end it this way than the other.

We hang up, and the entire reality of you swirls out of my kitchen and seals itself back into the phone like it's a genie lamp. I wish you were happy. I wish it didn't hurt either of us to lose each other. I know this will be the last time we talk.

I wish it wasn't so hard to say goodbye even when it is long overdue...

Integrity is my favorite quality in my favorite people, and the trait that I aspire to strengthen in myself.

The verb is integrate, the opposite is disintegrate. I will stop fragmenting myself, splitting between selves. I will learn to honor the truths of my high self, and not operate from the desperation of my wounding.

Integrity (*n.*): the quality of being whole and undivided.

I can't resist the urge to make patterns
Threading my way through
Days of green velvet meadows and
Draped shadows across suede dunes
at the edges of the world;
The pooled silk of tomorrows that beckon,
The lure of adventures and their homecomings.

My tracks on a map like lace
Woven flowers from central pistils
Radiating out and returning
Embroidering myself to this life,
I am living my art.

One way to ensure that you will have an interesting and adventurous life is simply to impulsively book non-refundable flight tickets... Then you just play the game of tending to the logistics as much as you can, because whatever else happens, you've got a flight to catch.

I am realizing that I don't want to be away all the time anymore. My reasons for traveling are changing, or perhaps one reason has passed: I finally feel like I belong to my life at home, and I no longer want to run away. But I find that I have still not eluded my wanderlust- the lure of exotic cultures tugs at my shirt hem, the gem of human life with its multiplicity of facets asks for me to come witness. Something arrived in the wind today, whispering my name. It was a day of animus and impetus.

After weeks of feeling miserably sick from the "cure", I threw my cancer meds in the trash and booked a flight to Morocco.

Canada

In the course of our lives, we will come to many tests that will ask us: *are you willing to grow?*

Sometimes we will not have a choice in our being delivered before we are ready to be reborn, circumstances around us will fall away and there remains no womb in which to stay secure and asleep, no way to go back and finish the dream. These times of great upheaval offer us access to the potency of reframing, renewing and rebuilding, starting from the place we fell... or remaining there. Whether we carve out a new life on this side of the portal is our choice, but regardless, we can never go back through it.

Sometimes, though, we can choose when to take a test. We can decide that we will surrender ourselves to a cocoon, to dissolve and permit our own reconstruction in order to continue on our path of leveling up in consciousness. Choosing tests gives us the option of being held through them. We can buttress our resilience by continuing our growth in times of peace. We can find or make a Container in which to fall apart so that our pieces don't scatter in the wind like petals. We can be deliberate.

In neither of these scenarios do we really know what awaits us on the other side of the trial. Courage is required in both to see things clearly and as they are, and to assimilate into our new forms only the pieces we want to incorporate.

If things are fine, why would you want to change anything? I feel this question in the ethers around me, sometimes from those who love me, and sometimes from within when I let fear in. I feel the soft edges of that sentiment, asking *why are things not enough as they are?* But the question, the wound, comes from a misunderstanding, not a rejection. Everything is perfect as it is. Each moment we experience leads to the next, and when we are awake, this journey is its own reward. Every single thing is in flux anyway, on its own course. We are the writers, assigning narratives to our collections of discrete and unrepeatable moments... but there is a great joy in letting go of all that we think we know.

I've spent the last few years of my life seeking, leveraging, trying to understand the nature of myself and the nature of the world. At times I was running away, too injured to go home, too vulnerable to be seen. And at times I was on the very cusp of my aliveness, blooming into a fullness of expanded potentials. Maybe both things were happening as parallel, concurrent truths. But nonetheless, I had to throw myself off of cliffs to believe in gravity. I had to throw my heart against things to understand what makes it shatter.

Now resilient, now home, now full of these wisdoms, the nature of the dynamism that I require to teach me is evolving, becoming more subtle. But I have not stopped the experiment, it has become how I live and how I love. I am adapting to new ways of opening, but I demand a life that opens me. Only when I am growing do I feel the essence of being alive.

Acceptance of change, and orchestrating the changes we want to embody within, is ours if we want it. The propulsion that drives me to continue to seek, to challenge myself to look harder, to improve myself even when I'm finally doing well, is not discontent or perfectionism- it lies in my eternal curiosity, in my innate desire to progress toward a higher self, to understand and connect. I am a dreamer, crossing back and forth between worlds in an effort to become a bridge between what is and what can be. I am a student and a child, and I want to experience as much as I can while I am here.

In two days, I will build a little cocoon, one of many I've known. It makes sense that we can only construct them from what is around us or from what we already are, and in that way, I feel better prepared than I ever have to undergo transformation. Whatever this has to teach me, I am patiently awaiting it, and I am listening.

Just two short days from now, I will undertake my first Vipassana Meditation course.

Conversations with myself.

"But I miss him."

"No, you miss the *moments* that you had. And those things are transient, temporary, already gone. Even if he was part of your life now, you would never get to repeat those same stories. You are imagining continuance where it does not exist, and this is how you get tangled. Your lives intersected but are not slated to continue in parallel. You are mourning *memories*- but you can revisit those memories on your own, they are yours to cherish. You are mourning the loss of potentials. But focus on the bliss that comes from being able to create new moments. Do not attach to moments. Nothing is lost that wasn't always lost. Understand that this is the nature of things so that it does not hurt you. You are attached to the way people make you feel sometimes- recognize that for what it is, and be gentle with yourself."

As I decouple these things, the lonely thought hits me that perhaps I've only been *addicted* to people, and so rarely ever in love outside of myself. Perhaps I have fallen in love with the way that I *felt* because of people, because of moments. I roll this around to test its truth. Many loves in my life have not been spared the poison of attachment, even as freely flowing as they seemed at points in time. I have called in others to love me when I could not love myself. I have found my face in theirs, aspects of myself manifested before me, asking what parts I could bear to see, what parts I was willing to forgive, what wounds I would tend, who I wanted to be. Sitting with this in aching, granular compassion, new truths unfold.

What is different about a love that transcends? The mirror is covered, the self looks outward, only wanting for the other. There is no desperation, no grip of longing. Only warmth. In my life, one relationship stands apart from the rest- the one that I have impetuously tried to sabotage into exposure of its falsity, but that waits with the patience of angels for me to understand that it remains untouchable, unconditional.

* * *

I crossed the border at Wild Horse, and I embodied her namesake. I broke the reins and bolted ahead, the rolling, empty land falling away in the rearview, a violet-gray wall of thunderheads like wolves gnashing at my heels.

I couldn't escape the weird feeling that I was driving to my own birth. I tried to waft off the expectations that flitted around like moths trying to land on a lamp. But that lamp was a lighthouse lit, and I was barreling toward it from out of the darkness of all that came before. Maybe these next ten days wouldn't change me, but everything inside me hummed with fluttering anticipation that they would.

It's as though I'm going in for a sort of self-surgery, to chrysalise the silk of Noble Silence and discipline so that they can wrap me as I dissolve and reintegrate. I don't know what will happen. I only know that I will show up tomorrow, climb the stairs of the pharos, and surrender myself to the process.

I've fallen apart a hundred times but never on purpose. It already feels restful this way, knowing that I can unbind my pages and lay them out inside, in the wind-sheltered harbor of the great halls. Everything is designed to allow me the time and safety to be intentional in my reintegration.

I think for a moment about the last time I dissolved, really and truly dissolved, the violence of it. He had erased things about me, rewritten some. When the hurricane hit the library of my inner selves, he held my arms. Nothing was spared. When I finally got free of him and the storm had passed, I managed to salvage a few tattered pages but I could no longer remember what the difference was between my handwriting and his. I was so barren at first, the stark white of pages utterly distrustful of committing to ink. It took so much time to rewrite myself, to learn who I was and what I valued enough to say with surety.

I can't imagine how blissful it will feel to unpack this time, held in Love. I am not afraid of anything I will find in there, I am the author of all of it. I don't fear the task of putting myself back together, I've had great practice for that by now. I am so ready for what this process will bring. I enter this place hopeful and open to metamorphosis.

Within the Four Walls of My Mind

Time does not exist here, only breath. Breath and the mind. Those bridges between the conscious and subconscious that we can direct with intention or let them do what they do. The subconscious, Animal, will dominate them if she is allowed to. Animal will "breathe you" whether you think about it or not. Shallow for panic, deep for calm. Animal thoughts are automatic and come from the conditioning, the environment, the chemical body responding. She is impulsive, biological, reactive. The mind will automatically busy itself, mired in sensory inputs, evaluation, superficiality... clinging, aversion, clinging, aversion. Animal is the default within the body, acting in the interest of survival. Separate now from her own wild nature, she is maladroit in the throes of modernity. She is the self that is unaware.

Here, we attempt her training, her marriage with awareness. She does not want to be tamed. She bucks against the restraints, fidgets, highjacks the mind and takes it for joyrides. In lavish escapades she imagines driving away, singing, loving, eating, running, traveling. She will not be made to sit still. Patiently, Soul leads her back again and again. We are doing this.

Now Soul observes the breath, the natural breath, as it comes and goes. Animal thrashes. She is pinned down but won't be pet.

But then, after some time, something happens. Timidly,

Animal emerges and surveys this strange scene, watches Soul watching her. And she permits it out of curiosity. By the end of Day 2, Animal leaves her eyes closed so that Soul can work. She does not yet understand this discipline we are attempting. But she is bored enough now to entertain it for a few minutes at a time.

The thing that can harm you the most is a wild, untamed mind. But the greatest ally you could make in your life is a mind where you are the master. It's as magical and elusive as controlling the weather. I look around the meditation hall. I have no idea the experience that the people around me are having, as we are in silence for ten whole days. We meditate more than ten hours a day, and they all look so calm and stoic. *Is this as hard as it feels?* Maybe they are training Golden Retriever Minds. I am attempting to harness a Wild Beast who has never before been touched, let alone ridden. A Raging Bull Mind, whites of the eyes showing on my every approach. This unbridled mind and I are stuck here together now, in supported but complete isolation. There is time to just sit with each other and be. *You be wild, and I will be patient,* I tell my mind. It's ok. We have to start where we are.

* * *

Time passes. An hour or a thousand lives. There is no way of knowing. Clocks don't measure what is happening here.

At times the practice is easy, and I can't wait to slip back into this exploration of my body's literal experience. But I go entirely mad at least a few times a day. I remind myself that to be here is a choice, it is not a prison or a punishment. To optimize our minds requires sacrifice. To lead our best lives, we must do the work required, walk the path, not just believe that the path exists.

* * *

We are training the mind to focus itself intensely within the body, and we are doing this to understand our absolute objective reality. We are not experiencing anything as we wish it was, only what is Real. With the outside stimuli shut off, we can feel the real feelings of our bodies, not the imposed sensory reactions that stem from the external world. We are peeling back layers of programming. We are observing ourselves, these sensations inside the body, with equanimity. We are training ourselves out of automatic responses by observing everything without reacting. This is how one becomes wiser: by practicing a technique to become free of clinging and aversion.

Evening discourses explain more: *Everything is impermanent. This is the universal law of nature. It applies within the body and in the apparent reality in which we live. Every single thing we experience is transient, ephemeral. Our bodies, our homes, our kitchen tables, rivers, mountains, the sun, the walls of the room we are in... every single thing is made up of particles and these particles are in decay. They change. Each moment is fleeting. Even as a candle burns, the flame may look like the same flame for the course of some time, but an infinite number of flames are being extinguished and replaced as it burns. The candle is different in every moment. This is the nature of things. This is the nature, too, of our reactions, our constructs, our perceptions, our joys and pains. Our loves, our sorrows. Nothing is exempt from the law that Every Single Thing Is Impermanent.*

The synchronicity makes me tear up. I have been understanding this for so long on some level beyond language in my heart... that it is all moments. That people are flows, and our relationships with them are a unidirectional evolution. Encapsulating the essence of an experience is the object of art, nostalgia, and communication, but it is all in flux, moving, living. The qualia of our subjective reality are unique to us, and nothing is directly repeatable or preservable. I am *so acutely emotionally aware* of how much I change, how rereading my writing always seems so deeply familiar yet not my own anymore. I knew all of the women who wrote this book. They are like my wild ancestor-selves, but I am none of them any longer.

I have felt the changing self so intensely these last few years that I thought I was losing my mind. Maybe I'm just becoming more aware of these universal truths.

This technique offers somatic understanding: to feel the reality of the body as it changes; to train the mind to be so subtle as to detect these particles as they manifest and die, sometimes so rapidly that it feels like vibration, and sometimes much more slowly and painfully; to feel the cellular nature of the body and its absolute aliveness, its changing, its decay in real time; and to experience this all with a neutral mind that only observes.

* * *

I find old receipts from a coffee shop and a pen in my backpack and indulgently break the rules by jotting down microscopic notes as my mind is blown apart. My heart twinges with both failure and a simultaneous compassion that I cannot yet let go all the way. I am a daughter of trauma with a residual distrust of my own reality, and writing the story gives it topography, landmarks. I am mapping the geography of my soul as I infiltrate westward into the unexplored reaches of it, only to find it already inhabited. I pause in recognition within this metaphor. Right relationship is paramount, permission is required, so I offer myself for consideration, I ask and I wait. *Maps are not the way things are found or remembered here*, I am told from somewhere, from within. But I don't know the other ways not to get lost yet. There is an extraordinary pull to grow into the depth of relationship that doesn't require the keeping, but that expansiveness is what I am here to learn- it is not yet fully embodied. I make a solemn promise to myself that I will brave the trust required to embrace the fullness of ephemerality next time, soon, and as much as I can moment by moment. Right now this feels like the only way I will make it through. A breadcrumb trail into the cosmos. The expansiveness is unexpected, even if you think you will be ready to meet it. I am downloading vastness into my person and unable to explain it- it's too big, too much, incomprehensible even as I feel it entirely. *Animal doesn't need language to know things*, I realize in a non-verbal pulse of sentience.

*   *   *

That we are ever in a solid state is an illusion. Tangible matter is simply energy that is vibrating slowly. Every particle of being is in its own life cycle, dying and being replaced constantly, continuously. Our mental and emotional states are the same way. What matters is who we are in this moment, in every moment as it is happening. *Can we be embodied in who, how, and where we arc? Can we inhabit the present instant as it is?* To internalize and deeply understand the impermanence of everything is to grasp the essence of reality. In disciplining the mind, these truths can hold us as everything falls apart and reconfigures over and over again.

*   *   *

I think back to the day I was sitting in my therapist's office-now over year ago- when she told me that my ex was but a symptom of my existential crisis. There was a seed of truth there, but it wasn't the whole truth. I see this so clearly now. All of the people in my life- my partner, my lovers, my friends, and even my family, are symptomatic of the expression of my soul. Every person in relationship with me is but another part of the same whole. We experience others as aspects of ourselves made manifest in the embodied dream of life by the soul coding that creates us. I came here with things to learn, karmic stories to play out, archetypes to encounter. There were templates for my experience encoded in my soul contract and I have lived them out, given them direct expression as was my destiny. This is the dance. As a result of my energetic imprint, these are the people that have arrived, called to me by the siren songs of my love, pain, and circumstance to show me what it means to be human. They have all been Me all along, and me, them. Each of us are encountering the interplay of individuation and oneness by looking into these mirrors of relationship.

I am not an existential crisis, as my therapist so insouciantly suggested. I am both so much more and so much less. I am the divine awakening into humanness. I am human awakening into divinity. I am witnessing my software and challenging it to yield to my conscious evolution. I feel the boundary of the contract, the limits of my person in the context of her life- but also the continuity of my being, and an opening into love for that leap of faith into the unknown between them.

And I am witnessing myself ask this of my human experience: let me ascend into Knowing, and let me be held whether or not I arrive there.

We are each but waves on an ocean, unique in expression, but never really apart from the whole. We are curlicues unfurling at the furthest peripheries of the Big Bang as it continues to expand, to paraphrase Watts. From the great sea of the unmanifest, we slow into matter and form, our energy drawing from the ethers and arriving into consequence. We are bound by our claim on time and space, by the epic poems of which we are but one line. We are evolution of a process in real time, at once inconsequential and profoundly significant. In the blink of an eye that is our physical embodiment, we are given the gift of this earthly experience- a consciousness that we use to both explore and transcend our interpretation of ourselves. *Let me be Love either way*, I whisper.

The recognition pours through me like light, loosening the hold of gravity between my cells and self-concept as I am delivered into an expansive belonging that defies the limits of my human faculties. I feel electrically connected to a web of life that includes the inanimate and mundane, the planet itself, the universe. What matters in my life is being screened before me, a movie reel of memories, of people I love. How I love them each so completely for the gifts they've given me, for defining the life that contains me. *Am I dying?* Gratitude shatters me in a lightning explosion through my body, and I drop to the floor in utter awe. I catch my breath and look at my hands, coming back into my body. Laughter washes over me in purifying waves, as I kneel alone on the cold tile in my dorm room, staring at the fronts and backs of my hands in wonderment at the absurd plainness of the miracle. If this is insanity, I prefer it completely.

* * *

Can we ever really transcend the four walls of the mind? Can we imagine and create a mind that exceeds the perceived potentials of the biological machine in which it resides? Can we surrender from the grasp of labels, expectations, fixations, and narratives that comprise the construct of our identity as we have built it? Can we dare to tap into the threads of a consciousness that holds us as parts of something much greater? We are but cells within a cosmic organism beyond the comprehension of our finite intelligence. We are multiple masks on the same character, unknown to ourselves.

* * *

I understand this technique inherently, it feels so natural to me to undergo this experiential learning. The process is a controlled practice, the breakthroughs in consciousness are spontaneous but just as real. I am tactile, I have to feel things to know them. It's the only learning that has ever made sense to me, but I have been chasing it in externalities. I've been using my senses to examine and define myself in terms of the world outside, and now I turn those faculties inward- not in some theoretical or philosophical exploration, but a perceptible and literal one. I am training my mind out of habit patterns that keep me in a cycle of superficiality and in the illusion of separateness. I feel more hope than I ever have for absolute healing and true happiness. I feel an unmistakable presence of being.

I flash back to the ruddy-faced voyeurs pressed against the bus windows in Peru. I don't want to be a tourist in my own life. Can I become strange enough to hold the strange truths that are opening before me? We are flying through space, and humans are so much more than human, and our consciousness is part of a much bigger consciousness. We are the ongoing and current expressions of a universe that is alive. And nothing, none of us is separate. We are not just from this world, we *are* this world.

I look at my life from outer space, and I see a girl falling in and out of love, tossing in the waves of her emotions... but then learning to broaden the desire that drives her, falling in love with being alive. The spark doesn't diminish in this shift, it's not a repression or a wholesome attempt at a moral containment. It's an awakening into the fullness of potency, tapping into the larger current of aliveness that compels us into manifestation and expression. I can *feel* the truth in this, I can feel the diaphanous filament of the electrical connections in flux between everything, holding us in a field of relationship. I can feel our sheer will to exist, and our misguided fears about dying. If only I could share how readily the illusion of material reality can be pulled back to reveal the wheels of the firmament.

How plainly the mystery sits before us, waiting for us to learn how to see.

*   *   *

I've accumulated a small secret stash of crumpled receipts full of my thoughts. I am partly ashamed that I couldn't avoid the temptation to document my experience, so I try harder to observe myself feeling this way without reacting. But I don't stop writing. I read the tiny scribbled notes and already they are the words of past selves. The present unravels faster than I could write it, and there is no way to compress it here. I wonder about the half-life of selves. Is it shorter now than ever because I am changing so much here, or has my awareness just become more acute?

*   *   *

There is an emotional component of the practice that I observe. I feel a profound sadness for actually *feeling* Animal dying. But also I feel a great relief, like I am comforting her by not turning away from this truth.

*   *   *

This is rehab for my passion addiction, learning to come out of the cycles of attachment and avulsion that are holding me down. I am in extreme withdrawal here but I am receptive. I have always been a reactive person, an expressive and sensitive person. And I see so clearly now that I didn't understand the difference between created vibrations and inherent, real vibrations from within. I have been bouncing off of externally-derived sensations, chasing highs to try to recover Feeling after the numbness of depression. But deep down, all I wanted was to experience reality, real sensation, truth. I wanted to heal and was doing all that I could to find truth *out there*. But the truth can be found right here, within, right now, any now. Layers are removed only to reveal more layers, and truths that defy language burst forth from the locus of my attention within.

I observe my mind being taken apart, I am able to literally *feel* the rewiring. The intensity is surreal. Dreams come by the handfuls: strange, lucid, dripping with archetypal context. Even in my dreams I am observing myself differently, challenging my reactivity. I dream of a tarot-based merry-go-round in Paris. I dream of a whale blowing water at me and telling me that she has often thought of me too. I remember at least four or five every time I wake up, but I don't use my limited scraps of paper to write them down.

Memories surface that I have not thought of since they happened. Some are random, but many, most, are pivotal moments that hurt me and redirected my life thereafter in subconscious avoidance. I remember my kindergarten bus driver inviting me to be the one to turn on the flashing lights for him one day. I remember the Valentine's Day classroom party in second grade being cut short because it was snowing. I remember my father shaming me for my extreme attachment to my cat when I was seven, as I was going out in the pouring rain to find him. I remember my grade school basketball coach driving me home one night, and lewdly telling me he'd noticed that I was more developed than the other girls in my class. I remember showing my Mom my self-harm wounds when I was fourteen and telling her I needed help, to which she replied, "We all have problems." I remember my boyfriend when I was 20 telling me that there was something remarkable about me, but I just wasn't conventionally beautiful.

Years of compounded shame for my body and my desires and my sensitivities break free, released of their hidden holds and all that kept them dammed in the channels of my being. Thousands of memories and their respective emotions boil up from the dark trenches of my mind, and I realize they are as relevant to observe objectively as the coming and going of sensations in the body. They are being freed from the *samskaras* and mental knots, back into the mindstream as they are released. They are testing my equanimity. I let them come, and I let them go. My reality is not in any past, it is right here and right now. My life is *truly* only happening right here, right now. After the flood has passed, I feel compelled to offer forgiveness, but only to the extent that it is genuine. I am full of a broad love and fire-forged resilience, but I still ache from realizing where my patterns of pain and maladaptation began. I am still reeling from the recognition of lineage trauma. It will take time.

Some of the memories are the good ones, moments of sheer bliss. These are the ones I have clung to, tried to repeat. I have sometimes allowed people to stay in my life hoping that we could recreate the exact times I felt such happiness, such validation. I have invented things within people that were not there. I have seen what I wanted to see, needed to see. I have made lovers into drugs, but many only got me high the first time. I didn't know I was using people, these patterns were buried so deeply within me. Of course, nothing was so reductive as it happened. I had no awareness of my reasons for choosing and allowing what I did, just a devastatingly porous willingness to Feel something. I reveled in the sweet self-destruction of losing myself in union. I really did love those who I have loved, want for their joy, want to give them all I could. But in my brokenness, I became infatuated with people who reflected me back in ways that I was starved for. And I loved that image of myself in them more than I loved them. How it hurts to see these things clearly. The pain of having treated others this way, having perpetuated these hidden injuries, gets stuck in my throat and is an actual knot. I spend multiple sessions sitting with it, observing my inability to scan quickly through such dense sensations. The memories are gone but they are stuck in my body, in my energy body. I am beginning to see the relationship, the reality.

Soul says we don't have to suffer this way anymore, we don't have to act out of desperation. We don't have to react at all now that we know this way. I can stop these patterns of misery in my life. I need not "work" on any memories or employ them to punish myself for my mistakes. I need only to be right here right now, objectively and compassionately. I am exhausted, but overall very deeply happy.

* * *

I am coming out of the woods now, there is more light between the trees. Everything was dark and obscured, and now a path has emerged. The work is still the same, regardless: merely placing each foot as it falls on the ground before me.

* * *

The mind again oscillates into rebellion, wants comfort, release. There is only more meditation.

A storm comes and I love it. I find a four leaf clover and it is a salve for the grit grinding within my soul that is making a pearl. It is a reminder that *I've got this*. No moment is the same as the last, and each has its own fruits to bear. I make up stories about what the birds' lives are like and intuit what they are saying to each other. Outside the window, bird drama. Inside: tea, meditation, discomfort, objectivity, surrender, repeat.

Somewhere, rooms away, a ventilation duct softly thumps and becomes a ceremonial drum assisting me in my scan. Obsessive thoughts about time. Dissolving. Dissolving apart. No thoughts, only feelings. No feelings, only body.

And then, all at once- no body. I vibrate apart in a complete lack of all obsession, a coming up for air, a ceasing to exist... *what IS this?!*

It's like I was so far inside my body that I was free of it entirely. But the feeling is gone in a flicker and I can't find it again, like a radio station just out of range.

*   *   *

For all the discomforts here, an undercurrent of peace is beginning to reveal itself. Here, I witness in real time, my own metamorphosis. When the panic comes, I return to observing breath.

*   *   *

Animal thrashes against the chains less frequently now. She is not giving up, but rather, she is coming to understand. The bondage she is in is not that of this place, these people, these rules... it is of her own making. It is the result of having an ignorant mind, but she is now awakening, she is being freed.

Everything feels different now, heightened, but not from the usual oscillation between passion and destruction. Things feel Real, and I feel them in my body.

Afternoons are the most challenging. Animal grows impatient and fights at every step. For every deeper level of comprehension, there comes an obstacle of equal caliber. I am clinging to ideas of going home.

*   *   *

I have been so proud of the ways I have been superficial and passionate. Proud of all the ways I could lose myself to the experience of my senses and find such entertainment in the drama of my circumstance. I had no idea the injury and injustice I was causing myself, how much I was the root of all my problems. Now I see that I am also the only potential solution to them.

How differently I see things now. How excited I was to be able to feel so much, to be moved so freely, to let these feelings carry me with them into whatever thoughts and actions that resulted from them. It's no wonder that I have suffered so badly.

I think for some time now, I have been able to feel things profoundly, both within the world and within myself, but I was attached to moments as though they would last, places as though they would be the same. The awareness was blooming, but I clung to every sensation and let nothing pass that didn't cause longing. Now the awareness is being honed, and I am discovering a peace I have never known in my life. I am letting go of literally Everything.

*   *   *

Lifetimes of thoughts, galaxies of thoughts. Perception as fractals. A cosmic river is raging here. When I can observe the river, I am not afraid. But when I fall in, I am quickly overwhelmed and struggling not to drown. I go back and forth, trying to access the bodily reassurance that I know how to swim.

*   *   *

I think that in all my travels, part of my impetus was a *seeking*, and what I was looking for was access to universal truths. Like perhaps if I could understand the world as it is and assimilate those wisdoms, I could finally make peace with myself. If I could just go and see everything for myself, touch everything, experience everything, then I would *know*, and my knowing would pull me into resonance. But my thinking was inside out. Experiencing the world *within* is the only way to arrive at the embodiment of real understanding. In our truth we know: everything changes, everything is impermanent, and we are calling to us that which aligns with our energetic frequency. If you don't already feel these truths from the inside, you will never find them in your surroundings, only an endless mirage thinly veiling your unhealed aspects, and a fruitless search for the medicine you already possess.

*   *   *

It is commonly accepted that happiness comes from within, happiness is a state of mind. But while people may appreciate that sentiment intellectually, no one tells you *how* to achieve this. The generation of happiness comes from breaking apart the stuck habit patterns of the mind. Any joys that arise from external influences are as fleeting as their source. Transcendent happiness is the freedom from attachment.

*   *   *

In the afternoon I experience an explosive total system failure of focus and attitude. Of course, this is completely contained within my head. *Everything here is contained within my head. Everything I have ever experienced is contained within my head.* I look around and wonder if anyone else is losing their shit. They all look so calm. Then again, probably so do I. I manage to return to focusing on the breath and eventually that pulls the wreck out of the ditch. *Ahhh, ok, I should remain balanced about this too.* I look around again. Nothing has changed. Nothing has changed and yet everything has changed. We live our entire lives inside of these limited and obsessive little minds that flutter like the leaves of an aspen. It is an enormous undertaking to train one.

* * *

Flow during a scan is like dissolution of your entire self-structure into a trillion little tingling particles that suddenly have all this space between them. It is weightless and full of light. We are reminded to be careful not to crave this, but it is truly incredible when it happens. To think that we are only allowed to observe reality, to know that this sensation is Real and resulting from the ability of the mind to focus and observe... it is beyond all comprehension or expression. It must be witnessed individually to be believed. No words, nor the five senses, can contain it.

* * *

I am, and I have been, a person full of love. But I am feeling clearly how much of that love has been limited by attachment and desperation. Every time a love fell apart, the entire world was ending for me. I have thought a lot about how I never gained a sense of belonging or secure attachment from my family as a child. I ache for the aches of my parents and ancestors, for that which they could not resolve, that pain that they both inherited and passed on. I ache for the ways they were not able to awaken, and I humbly and completely forgive them. I take responsibility for myself now, carrying only what is mine. I know I can heal this broken aspect of myself and continue to purify the way that I love. It hasn't *all* been mistakes, it has been so perfectly human. The process of healing has already shifted the axis of my whole world. Still, I reread my stories and I know how much work there is yet to do when it comes to love... falling in love, being in love, being loved, loving, losing love, losing loved ones...

* * *

There is no anesthetic here as the knife cuts even deeper, but I know this operation will save my life.

* * *

Day 8. One more hard day after this. I try to stop obsessing about time. Or I at least observe myself in the state of obsession. I have lived ten thousand lives in here.

Suspended self, expanded self. Both at once. Neither at once. Quantum selves. I feel my brain unlocking, unwiring, the knots unraveling. I see myself from just outside my body, I feel everything expansively, as though the blind spots in my peripheral have been pulled into view- curtains flung open to reveal that the whole wall is actually windows. It's terrific in all the complex senses of that word. Terrifying, exhilarating, mind-blowing... but there is little time to capture or reflect on the process, the revelation happens faster than I can attach words or thoughts to it. I'm laid bare by this new comprehension of my own infinite inner space. A space with which I've always had familiarity, but that I now feel in carbons and neons, synesthetically as though my whole brain is open and my senses overlap to assist each other in understanding the vastness and reality of it. Language fails me completely. I am made of pinpricks of light. I am made of moments. Forever is made of moments. I am made of forever.

Soul and Animal have started to marry, I have started to become Real. Sacred Animal, Awakened Animal, Temporary Animal. Transcendent animal. *Real*-ization.

* * *

By the end of my time here, I no longer count the laps I walk in the yard at lunch, or the hours until sleep, or the days until home, but I am unsure if I am overcoming this or if there is simply nothing left to count. Being here in isolation challenges my deepest wound: my fear of abandonment. I miss Andy and Loki so much that I struggle at times to be present. I am still suffering, but it seems to lessen with the more I meditate.

* * *

I have lived my life as a hedonist, a seeker of pleasure, a numb-er of pain. I have been a slave to sensation. I have only been capable of distracting myself from this affliction from time to time, but never of curing it at its root. The work I am now doing has the potential to free me, I can feel this as it happens. Gratification cannot perpetually insulate us from impermanence, nor can it extricate us from inadequacy into enoughness. We will suffer losses, we will change, we will grieve, and how we are able to show up for that is a measure of our true capacity, our willingness to drink thirstily from the cup of all that is our humanity. In a similar vein, intellectual wisdom is profoundly enjoyable and has utility in the material world, but it can't usher us into presence within our experience. No amount of knowledge can teach us how to *be* within our moments. There is a difference between reading about swimming and jumping into the ocean.

There is no replacement for the experiential life, and there is a distinction being born within me between what that means from a sensory perspective and what it means from a witness point of view. No sensation or thought can carry the same magnitude of wisdom as that derived from beholding the evanescent state of our reality from within our body. Taking a deeper seat within ourselves can ease the splitting affect of our dualistic constructs as it reveals to us a nature- our nature- made of cycles and circles, and a wholeness from which we are never cleaved.

* * *

I can feel the quality of being alive, of being Real, of dying. I can feel the illusion of solidity and the reality of change. It's electrical.

* * *

I had my most mind-bending meditation ever this morning, and then later I went completely crazy. I wanted to drive away or at least sneak out to the car to get my phone. The impulse was so severe that I couldn't mentally sweep my body in meditation, couldn't scan any part of it, couldn't return to breath. They say these types of tests arise when we are ready to face them. But all I could do was *barely* prevent my physical action. I could not stop the thoughts by coming back to the body. I laid in bed in my bare room suffering until I was finally able to see that these are the exact challenges that I came here to conquer. No one here is hurting me but me. No one here is the enemy. The enemy is within, and submitting to my complexes only strengthens them. Even if all I can do is stay put, I am still being healed. I'm like an errant child, the urge is relentless. I realize that I can leverage this situation for progress. Taking a step back, I can be compassionate with myself and I can be equanimous with my condition. Slowly, slowly, slowly, I come back to breath. Sweeping the body is like dragging a long-toothed rake through a swamp. *There is just as much to learn from observing this as any other reality that I could be experiencing*. These extremes are the process of purification. Sooner or later the discipline will come, and these impulses will die before they surface into my conscious mind. For now, I can only continue to try from right where I am.

Stay here and I will pet you with my attention, Soul whispers. Animal's heart races. She is not used to being touched.

* * *

I sit in the shade of a crooked tree in the heat of the

afternoon. There is a soft rustle of wind through the tippy tops of tall cottonwood trees, and the hum of a bumblebee in the clover. The wind entrains seeds in the thick air and lazily carries them to their respective somewheres. How natural it has become to just watch things now.

*   *   *

Day 10. I'm trying to observe my elation and not fall in. Not change it, but not let it overpower me.

Reentry into the real world comes in phases so that we can adjust. Today we learned Metta, a meditation that sends out love and compassion to every living thing. *May all beings be happy. May all beings be peaceful. May all beings be liberated from their suffering.* My entrainment in this frequency is absolute; I am vibrating so high I can barely stay in my skin. And yet, it's the only place I really am.

Now, we are allowed to talk again. We come out of Noble Silence as though we are in post-operative delirium. We are ecstatic and full of love, but weird and sensitive. We sit together outside in a circle below the crooked tree and finally get to offer each other the compliments we've been holding in. We talk about our experiences and listen with immense curiosity as others share theirs. It feels so soothing to laugh. The grass is blindingly green. Words are careful, or too loud. Everything is Real. I have to take a break and take a shower. My heart is racing. In the shower, warm water pouring over me, I weep in gratitude.

At night, we pile into one of the dorm rooms, all twenty-five of us women propped on pillows and sandwiched happily onto cots. We are already bonded from the parallel potency of our individual journeys, and we cement our kindredship with laughter and a shared contraband bag of honey cough drops.

*   *   *

Day 11. I finish packing my backpack and finish the last cleaning details for my sparse little room that will soon host another woman through this process. I wish for her success and her strength. This has been the most extraordinary and challenging undertaking of my life.

I attend the closing chanting and discourse and say my heartfelt goodbyes. Today is the day of our release back into the wild.

Despite the sensitivity and sensory shock from being able to talk and hug and use the phone again, my inner voice sounds different to me now. I can see where this is all heading. You start to live a life behind the emotions instead of enmeshed in them. I'm so used to an existence of extreme excitements and extreme disappointments. If I can carry what I've witnessed here with me into my days, it promises to deliver me into a better way of life. I've been a self-deceiver, a great justifier of my own bullshit, and a mercurial emotional disaster at times... but I've also never shied away from growth. I've been a seeker of the truth, a healer for myself and others, and I've always been a person who has loved as much and as hard as I knew how. I am, and have been, a person evolving. I am glad to do all of this again and again, as many times as it takes to learn and re-learn non-attachment, right relationship, accountability, integrity, ease, peace, self-love, and universal truth. I am so, so full.

I have understood so many of these wisdoms intellectually, and occasionally I have felt their reality. But I've never been consistently able to align my life choices with the things I knew to be true. I was mired in cognitive dissonance, excuses, numbing, and distractions. I was conditioned by my pain, always running toward or away from something. So passionate, yet so committed to my own pattern of injury.

I knew I wanted to be happy, but I could never control my emotions. What a relief to finally understand that the way out of misery has nothing to do with control, and everything to do with training. It is not by *will*, but by practice that the mind can experience reality clearly, witnessing without evaluation our transient roles as embodied and conscious beings, with all the inherent meaning our mortality lends to our time here.

The very essence of our aliveness lies in the gap of potential, in the synapse of possibility. Liminal secrets swim in the energy of our interconnection. The fog of my spiritual and mental afflictions is lifting in the warmth of the sun. I have never felt more like I belong with this body, in this place in time and space.

Happiness is being only and exactly where you are. I wonder how often I will find myself there.

I mean... *Here*.

I point my car south and start the journey home. I have a long way to go.

A crew of ancestral selves at the helm
Steering us all into the edge of presence,
The birth of brave new versions
And the quenched flames of old ones.

Pushing onward
Delivered into moments anew,
Rising and falling in the chest of the sea;
Eyes shimmering like
half-spent coals raked into a pile,
Glistering embers recounting the combustion
into respective forevers
As each wave ushers us to the arrival of
now after now.

One by one
I awaken my selves and let them go
Embodying the extent of my aliveness.

I am where I am
Because of the choices of my own predecessors;
I am who I am
In an evolution unfurling-
My ascension into an awareness
that I now stand on shoulders,
Piling higher by the moment
as I shed spent selves like emptied armor.

Climbing upon them so that I can see further,
Perched now in the crows' nest
Looking out across the flickering fractured mirror,
Sun spilled like diamonds.

It doesn't matter where we are going.
It matters
That I am finally Here.

About the Author

Jen Baranovic is an embodiment and resilience facilitator, experience curator, ritual healing practitioner, death midwife, and writer. She is the descendent of river people and farmers, royalty and poverty, generations of impassioned wanderers and yarn-spinners. She was raised by wolves in the swamplands of southeast Missouri and now resides in the foothills of the Beartooth Mountains. She has led a thousand lives rolled into  one, and from the raw fodder born of misfortune and callow misadventure, she annealed her heart and honed her purpose in service. Her accounts of Becoming are that of a conscious evolution which holds reverent compassion for those selves we have been before.

Her writing is often gritty and vulnerable, exploring themes of personal mythology, belonging, loss, nostalgia, upheaval, and coming into authenticity. Her other books include *The Ways We Lost* and *Honoring the Sacred Wild*. She hosts retreats and community workshops that draw from a combination of healing modalities to lead others into connection, resonance, storytelling, and integrated aliveness.

Contact: singingbowlwellness@gmail.com

A portion of the profits from her work benefit organizations that support those who have endured trauma, mental illness, and loss.

Made in the USA
Middletown, DE
17 November 2025